Anaerobic Biological
Treatment Processes

Anaerobic Biological Treatment Processes

A symposium sponsored by
the Division of Water,
Air, and Waste Chemistry
at the 159th Meeting of
the American Chemical
Society, Houston, Tex.,
Feb. 26, 1970.

Frederick G. Pohland,

Symposium Chairman

ADVANCES IN CHEMISTRY SERIES **105**

AMERICAN CHEMICAL SOCIETY

WASHINGTON, D. C.   1971

Coden: ADCSHA

Library of Congress Catalog Card 74–176092

ISBN 8412–0131–5

PRINTED IN THE UNITED STATES OF AMERICA

Advances in Chemistry Series

Robert F. Gould, *Editor*

FOREWORD

ADVANCES IN CHEMISTRY SERIES was founded in 1949 by the American Chemical Society as an outlet for symposia and collections of data in special areas of topical interest that could not be accommodated in the Society's journals. It provides a medium for symposia that would otherwise be fragmented, their papers distributed among several journals or not published at all. Papers are refereed critically according to ACS editorial standards and receive the careful attention and processing characteristic of ACS publications. Papers published in ADVANCES IN CHEMISTRY SERIES are original contributions not published elsewhere in whole or major part and include reports of research as well as reviews since symposia may embrace both types of presentation.

CONTENTS

PREFACE

Anaerobic decomposition is a biologically mediated process indigenous to nature and capable of being simulated for treating wastes emanating from municipal, agricultural, and industrial activities. The process plays a prominent role in conventional waste treatment practices and often constitutes the largest single investment at water pollution control plants. Unfortunately, reports on the application of anaerobic processes for waste treatment have indicated difficulties with process stability which have resulted in a growing pessimism concerning the future value of anaerobic treatment to water pollution control. Indeed, there are some who consider anaerobic processes for waste treatment passé, eventually to be replaced by aerobic and/or physical-chemical methods, and that further study and emphasis on continued application may be futile or at least somewhat less than profitable.

In view of this seemingly emerging negative attitude, why then devote a volume of papers specifically to topics on anaerobic treatment processes? What is to be gained by a review and iteration of past experiences, some of which have been less than encouraging? Can the often rather vaguely appreciated occurrence of anaerobic metabolism in animals and its existence in natural ecosystems—*viz.*, marshes, bottom deposits of lakes and rivers—continue to serve as a sufficient justification for additional research and exploration? What characteristics of the process make definition and control so difficult and future application suspect? Finally, what does the future hold?

Some or perhaps all of these questions have intrigued if not baffled scientists and engineers since anaerobic metabolism was first characterized and applied to waste treatment. Integrated knowledge assembled from various disciplines is required to resolve these questions together with a communication of this knowledge between the scientists and engineers as well as the technicians often responsible for process design, operation, and control. Accordingly, the papers in this volume were contributed as a result of a symposium organized to review the state of the knowledge on anaerobic processes, to provide a forum for discussion and critique, and to delineate current progress and areas of potential development.

Consonant with these objectives, the authors of these papers were chosen specifically for their expertise and recognized contributions on anaerobic processes to the biological sciences and engineering fields. The volume should serve as a vehicle and stimulus for information exchange on a broad spectrum of topics commencing with the fundamental biological and chemical principles of anaerobic metabolism and progressing through specific analytical interpretations and techniques to engineering concepts and their applications to the design, operation, and control of anaerobic biological waste treatment processes.

Atlanta, Ga.   FREDERICK G. POHLAND
July 1971

1

The Methane Fermentations

MARTIN J. PINE

Department of Experimental Therapeutics, Roswell Park Memorial Institute,
New York State Department of Health, 666 Elm St., Buffalo, N. Y. 14203

*The anaerobic biological processes that produce methane
are reviewed. The organisms catalyzing these reactions are
a morphologically diverse collection of obligate anaerobes.
Commensal and symbiotic with other anaerobes, they reduce
or dismutate H_2, formate, acetate, methanol, and other fermentation products to CO_2 and methane, completing the
anaerobic cycle. Five species have been isolated and characterized so far. Methane is generated mainly either from
CO_2 or from substrate methyl groups, probably via methyl
vitamin B_{12} derivatives.*

Scientific interest in the gases produced by decomposing animal and
vegetable materials goes back to the early scientific revolution in the
writings of Robert Boyle and his assistant Denis Papin in 1682 and of
Stephen Hales in 1727 (1). The microbial flora responsible for biodegrada-
tion, in general, form many interdependent associations of species whose
development may be greatly determined initially by the substrates to be
attacked. Aerobically a wide variety of biological and synthetic compounds of aliphatic and cyclic character can be decomposed completely
by individual flora or sometimes by a succession of two organisms. The
anaerobic cycle is almost as competent in this purpose (2) except for the
absence of a pathway of net hydrocarbon decomposition. Also a more
complex association of at least two or even three or four successive fermentations by different flora are required for completion. At the end, a
comparatively small number of common fermentation products are converted to CO_2 and CH_4 and return to the aerobic cycle as marsh gas.
Volta is given credit for first identifying this gas as a distinct entity in
1776. He found "inflammable air" to be generated everywhere in the
neighborhood of decomposing vegetation in bodies of water and in the
soil (3). Later efforts in the era of modern chemistry were devoted to
identifying methane as the common combustible component of marsh

gas, the fire damp of coal mines and rumen gas, and to establishing its origin in microbial metabolism (3, 4).

General Characteristics of the Methane Bacteria

In common terminology, the terms methane bacteria and methanogenic or methane fermentations apply to the anaerobes that produce methane and to their individual reactions, respectively. The distinction should be borne in mind between these organisms and their reactions and those of unrelated bacteria of the aerobic cycle that consume methane. The methane bacteria occur as sarcinae, rods, and cocci. A most unusual representative, *Methanococcus vanniellii* is highly motile and has very fragile walls and large cell forms (5). The morphological diversity among the methane bacteria suggests either very great evolutionary divergence of individual species or even convergence from different origins. There should thus be no great expectation of a common physiology among the methane bacteria; nevertheless they do have a number of similar attributes. They are obligate anaerobes with great sensitivity to oxygen, which is one reason they are so difficult to isolate. Their substrate requirements appear simple and narrow. Thus at one near autotrophic extreme *M. vanniellii* grows on formate as a sole carbon and energy source and utilizes no additional substrate but H_2 (5). For some species, factors may be required that are peculiar to the immediate environment such as is provided in rumen fluid (6, 7) and may not be found in other, more common natural sources. The methane bacteria can grow over a wide range of temperatures (8). As a group they can be especially critical in the completion of anaerobic waste disposal (9). They grow only near neutrality, which accounts for the preservative action of primary fermentations (alcoholic, lactic, propionic) that originate in acid media or that produce acid and for the problems resulting from acid production or acid contamination in wastes. The sensitivity to certain chlorinated hydrocarbons (10) provides one example of selective toxicity towards these flora that might be examined in industrial wastes. Methane bacteria have been most recently studied for their role, along with other organisms, in the production of toxic methyl mercurials from waste mercury (11). A spectacular episode of human poisoning from the consumption of fish that concentrate the mercurials (12) has brought to public attention considerable concern for contamination of the environment by mercury. No doubt the role of sewage bacteria in the favorable and unfavorable alteration of toxic wastes will become more extensively investigated in unravelling the many ecological derangements resulting from water pollution.

Individual Methane Fermentations and Species

A partial and abbreviated scheme showing the interrelationship between the methane bacteria and other representatives of the anaerobic
carbon cycle is listed in Figure 1. The heavy arrows indicate methane
fermentations by individual species or perhaps in some cases by closely
dependent symbiotes. The remaining reactions of Figure 1 are catalyzed
by propionibacteria, clostridia, butyribacteria, and other anaerobes. The
general references of Wood (*13*), Barker (*4*), and Stadtman (*14*) may
be consulted for further details and additional fermentations.

Béchamp (*15*) in 1867 was the first to describe methane production
from a simple fermentation product, ethanol, and to attribute it to a microbial fermentation. The second product he found to be formed from
ethanol was caproate, now known to be produced by *Clostridium kluyveri*
(*4*). Thus, carbon–carbon bonds are not only destroyed early in the
fermentation chain, they can also be reductively synthesized if the reaction is paired with the energy yielding oxidation of another substrate (*cf.*
initial fermentations of acetate and ethanol and also of H_2, Figure 1).
Several successive fermentations may then be required to convert these
products to methane and CO_2.

Hoppe-Seyler (*16*) in 1887 in his studies of cellulose enrichment
bacteria was the first to demonstrate a distinct methane fermentation,
using acetate as a substrate:

$$CH_3CO_2H \rightarrow CH_4 + CO_2 \tag{1}$$

The significance of the ethanol and acetate fermentations remained
in abeyance until Söhngen discovered the production of methane from
H_2 in 1906 (*17*).

$$4H_2 + CO_2 \rightarrow 2H_2O + CH_4 \tag{2}$$

He had found, as had Omelianski previously, that enrichment flora
growing anaerobically on cellulose normally produce CH_4 as an end
product but then usually produce H_2 after pasteurization. Isolated methane bacteria, as is now known, rarely or perhaps never sporulate, and of
the major flora that initially attack cellulose anaerobically, only the clostridia would have survived pasteurization. These organisms are vigorous
H_2 producers. Sohngen correctly reasoned that their H_2, together with
the $CaCO_3$ which was provided in the culture as a neutralizing agent,
were the immediate sources of methane, and he then directly demonstrated the reaction. He also isolated an aerobe that grew on methane
and oxidized it, establishing the final disposition of the products of the
anaerobic carbon cycle. Reaction 2 is the only fermentation common to
all methane bacteria. However, complete autotrophy on H_2 and CO_2

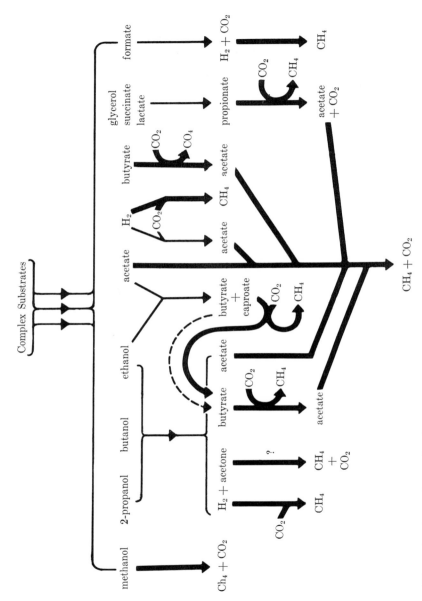

Figure 1.   Interrelationship between the methane bacteria and other substances of the anaerobic carbon cycle

without the requirement of minor organic factors has not been demonstrated unequivocally so far in any species. On the basis of Reaction 2, Barker and van Niel (*18*) generalized the methane fermentations as oxidation–reduction reactions in which the energy yielding oxidations of many substrates may be balanced by the reduction of CO_2 to CH_4. A number of oxidizable substrates can in theory replace H_2 as the CO_2 reductant, and several partial oxidations of organic substrates have been shown to require an external supply of CO_2 to be converted to CH_4. The oxidations include simple dehydrogenation (Reaction 3) and β and α decarboxylations (Reactions 4 and 5, respectively):

$$2C_2H_5OH + {}^{14}CO_2 \rightarrow 2CH_3CO_2H + {}^{14}CH_4 \ (18) \tag{3}$$

$$2CH_3(CH_2)_{(0-2)}CH_2CH_2CO_2H + 2H_2O + {}^{14}CO_2 \rightarrow$$
$$2CH_3(CH_2)_{(0-2)}CO_2H + 2CH_3CO_2H + {}^{14}CH_4 \ (19) \tag{4}$$

$$4CH_3CH_2CO_2H + 2H_2O + 3{}^{14}CO_2 \rightarrow$$
$$4CH_3CO_2H + 4CO_2 + 3{}^{14}CH_4 \ (19) \tag{5}$$

Reaction 3, attributed originally to a single organism *Methanobacillus omelianskii* (*18*) is the sum of the reactions of two symbionts (*6*):

$$2H_2O + 2C_2H_5OH \leftrightarrows 2CH_3CO_2H + 4H_2 \tag{6}$$

$$4H_2 + CO_2 \rightarrow H_2O + CH_4 \tag{2}$$

Upon cultivation on H_2 and CO_2, the non-methanogenic organism is lost. The methanogenic organism grows on no other substate than H_2. Reaction 6 is remarkable in being a putative energy source for the non-methanogenic symbiont, yet it is essentially freely reversible. The methanogenic component facilitates this reaction by continually removing product. The symbiotic system also fixes atmospheric nitrogen (*20*). Although Reactions 4 and 5 have been attributed to single organisms (*19*), the cultures were not purified, and they too might be symbiotic.

An additional two-step fermentation, catalyzed, however, by pure cultures of methane organisms is the decomposition of CO (*21*).

$$4H_2O + 4CO \rightarrow 4H_2 + 4CO_2 \tag{7}$$

$$4H_2 + CO_2 \rightarrow 2H_2O + CH_4 \tag{2}$$

The individual steps have not been discretely separated.

The methane fermentation of formate, for example, by *M. vanniellii* (*5*) must also to some extent involve two components—a hydrogenlyase reaction which can be the faster step and a methane fermentation of H_2:

$$4HCO_2H \rightarrow 4H_2 + 4CO_2 \tag{8}$$

$$4H_2 + CO_2 \rightarrow CH_4 + 2H_2O \tag{2}$$

$$\text{net: } 4HCO_2H \rightarrow CH_4 + 3CO_2 + 2H_2O \tag{9}$$

Also, the methane fermentation of formate by *M. omelianskii* is almost completely abolished by hypophosphite (*22*), a selective inhibitor of the formic hydrogenylase of the non-methanogenic symbiont (*23*). However, in the residual fermentation by the methanogenic component, some transfer of deuterium from the substrate to methane indicates a very limited amount of direct reduction:

$$4DCO_2H \rightarrow CH_3D + 3CO_2 + D_2O + HDO \text{ (22)} \tag{10}$$

In crude enrichment cultures fermenting formate, $^{14}CO_2$ sometimes does not incorporate to a major extent into the methane (*24*). It remains to be confirmed with the use of deuterium labeling and hypophosphite inhibition, whether direct formate reduction can occur significantly in pure cultures.

In summary the methane fermentation of H_2 (Reaction 2) is the only example so far which unequivocally uses CO_2 as a methane precursor, and it may be the only methanogenic component in most other substrate decompositions reported as methane fermentations. There are, however, two examples of methane fermentations where the major methane precursor is never CO_2 but an intact methyl group. In the methane fermentations of acetate and methanol, isotopic labels of the methyl groups are transferred without loss or randomization of their hydrogen substituents, to methane:

$$*CD_3{}^+CO_2H \rightarrow *CD_3H + {}^+CO_2 \text{ (25, 26)} \tag{11}$$

$$4*CD_3OH \rightarrow 3*CD_3H + *CO_2 + D_2O + HDO$$
$$(27 \text{ and unpublished results}) \tag{12}$$

The superscripts indicate separate labeling experiments. In neither of these two fermentations does externally supplied $^{14}CO_2$ significantly incorporate into methane.

To date only five species of methane bacteria have been obtained in pure culture. *M. vanniellii* (*5*), *Methanobacterium ruminantium* (*7*), *Methanobacterium mobilis* (*28*), and *Methanobacterium formicicum* (*21*) grow on formate and ferment H_2. The last species also ferments CO. The methanogenic symbiont of *M. omelianskii*, probably a variety of *M. formicicum* (*29*) grows only on H_2 (*16*). *Methanosarcina barkeri* is iso-

lated on methanol, grows less well on acetate, and ferments H_2 and CO
(21). This organism can, therefore, effectively use both the CO_2 reduction and methyl reduction pathways for methane production. The substrates of all these organisms appear extremely limited, but this is not
unusual among anaerobes. Only a small number of species of methane
bacteria may be needed in the environment. In the rumen of herbivores
the major methane organism whose growth can easily keep pace with the
turnover of the rumen contents is *M. ruminantium* (7) which ferments
formate and H_2. Formate and H_2 are thus the only early products of
cellulose fermentation that vigorously generate methane *in vitro* from
rumen contents (30), and the other products that could otherwise be
decomposed in methane fermentations are assimilated by the host. *M.
ruminantium* may be of advantage to the host in that the microbe's substrates are not of material general use, the considerable volume of H_2
otherwise eructated or belched is reduced, and the assimilated microbial
cell material is eventually digested (31).

Intermediates in the Methane Fermentation

The possible relationship of the methane fermentation with the more
conventional examples of one-carbon metabolism as catalyzed by folate
and vitamin B_{12} cofactors has been long apparent. 5-Methyl tetrahydrofolate, 5,10-methylene tetrahydrofolate, and methyl vitamin B_{12} are converted to methane by cell-free extracts of *M. barkeri* (32) and *M. omelianskii* (33). The involvement of vitamin B_{12} is further implicated by its
high cellular level in methane bacteria and by the isolation of B_{12}-
containing proteins in extracts of *M. barkeri* (30) which stimulate methane evolution from methyl vitamin B_{12}. The components and pathways
that can be demonstrated in cell-free *M. barkeri* extracts (32) are listed
below.

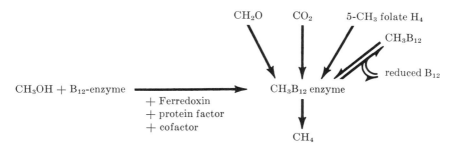

The involvement of folate in the methane fermentation is less clear.
5-Methyltetrahydrofolate is used only moderately as a methane source by
extracts of *M. barkeri* (14). No folate-containing factor has been demon-

strated in any methanogenic extracts, and indeed the folate content of the symbiotic *M. omeilianskii* system is low, on the order of that found in *E. coli* (unpublished results). The folate derivatives 5,10-methylene and 5-methyltetrahydrofolate are effective methane precursors with extracts of *M. omelianskii,* but they tend to be oxidized rather than reduced unless ATP is present (*32*).

The over-all reduction of CO_2 to CH_4 is expected to be a spontaneous process that goes through the reduction levels of formate, formaldehyde, and methanol with only a limited, perhaps early requirement for activation by ATP. At the lowest reduction stages, extra ATP may even be generated. In the fermentation of methanol by *M. barkeri* which utilizes only the last reduction step for methane formation (Reaction 12) somewhat more than 1 mole of ATP appears to be generated for each mole of CH_3OH oxidized to CO_2, judging from cell yields (*14*).

Our present understanding of the physiology of the methane bacteria is still in a comparatively early stage of development. However, pure cultures date back only to 1947 (*21*) and knowledge of their intermediary metabolism has depended on the comparatively recent elucidations of the active intermediates in the transfer of one-carbon units.

Improvement in the mass cultivation of representative organisms would greatly aid future investigations. A procedure for large scale cultivation of H_2-utilizing methane fermenters has been developed (*34*). *M. vannielii* is unique among pure cultures in growing rapidly and in good yield in a simple medium (*5*) and should be the organism of choice in the preparation of cell material.

Literature Cited

(1) Hales, S., "Vegetable Staticks (1727)," Oldbourne Science Library, Jarrold and Sons, Norwitch, England, 1961.
(2) Tarvin, D., Buswell, A. M., "The Methane Fermentation of Organic Acids and Carbohydrates," *J. Amer. Chem. Soc.* (1934) **56**, 1751–1755.
(3) Hoppe-Seyler, F., "Über die Gährung der Cellulose mit Bildung von Methan und Kohlensäure. 2. Der Zerfall der Cellulose durch Gährung unter Bildung von Methan und Kohlensaüre und die Ehrscheinungen, welche dieser Process veranlasst," *Z. Physiol. Chem.* (1886) **10**, 401–440.
(4) Barker, H. A., "Bacterial Fermentations," Wiley, New York, 1956.
(5) Stadtman, T. C., Barker, H. A., "Studies on the Methane Fermentation. X. A New Formate-Decomposing Bacterium, *Methanococcus vannielii,*" *J. Bacteriol.* (1951) **62**, 269–280.
(6) Bryant, M. P., Wolin, E. A., Wolin, M. J., Wolfe, R. S., "*Methanobacillus omelianskii,* A Symbiotic Association of Two Species of Bacteria," *Arch. Mikrobiol.* (1967) **59**, 20–31.
(7) Smith, P. H., Hungate, R. E., "Isolation and Characterization of *Methanobacterium ruminantium* n.sp.," *J. Bacteriol.* (1958) **75**, 713–718.
(8) Coolhaas, C., "Zur kenntnis der Dissimilation Fettsaurer," *Cent. Bakt.* (1928) **II 75**, 161–170.

(9) Gaudy, A. F., Gaudy, E. T., "Microbiology of Waste Waters," *Ann. Rev. Microbiol.* (1966) **20**, 319–336.

(10) Bauchop, T., "Inhibition of Rumen Methanogenesis by Methane Analogs," *J. Bacteriol.* (1967) **94**, 171–175.

(11) Wood, J. M., Kennedy, F. S., Rosen, C. G., "Synthesis of Methyl Mercury Compounds by Extracts of a Methanogenic Bacterium," *Nature* (1968) **220**, 173–174.

(12) Clarkson, T. W., "Biochemical Aspects of Mercury Poisoning," *J. Occupational Med.* (1968) **10**, 351–355.

(13) Wood, W. A., "Fermentation of Carbohydrates and Related Compounds," in "The Bacteria," I. C. Gunsalus, R. Y. Stanier, Eds., pp. 59–150, Academic, New York.

(14) Stadtman, T. C., "Methane Fermentation," *Ann. Rev. Microbiol.* (1967) **21**, 121–142.

(15) Béchamp, A., "Lettre à M. Dumas," *Ann. Chim. Phys.* (1868) Ser. 4, **13**, 103–111.

(16) Hoppe-Seyler, F., "Die Methangährung der Essigsäure," *Z. Phys. Chem.* (1887) **11**, 561–568.

(17) Söhngen, M. N. L., "Sur le Role du Methane dans la Vie Organique," *Rec. Trav. Chim.* (1906) **29**, 238–273.

(18) Barker, H. A., "On the Biochemistry of the Methane Fermentation," *Arch. Mikrobiol.* (1936) **7**, 404–419.

(19) Stadtman, T. C., Barker, H. A., "Studies on the Methane Fermentation. VIII. Tracer Experiments on Fatty Acid Oxidation by Methane Bacteria," *J. Bacteriol.* (1951) **61**, 67–80.

(20) Pine, M. J., Barker, H. A., "Studies on the Methane Bacteria. XI. Fixation of Atmospheric Nitrogen by *Methanobacillus omelianskii*," *J. Bacteriol* (1954) **68**, 589–591.

(21) Kluyver, A. J., Schnellen, C. G. T. P., "On the Fermentation of Carbon Monoxide by Pure Cultures of Methane Bacteria," *Arch. Biochem.* (1947) **14**, 57–70.

(22) Pine, M. J., "Methane Fermentation of Formate by *Methanobacillus omelianskii*," *J. Bacteriol.* (1958) **75**, 356–359.

(23) Reddy, C. A., Bryant, M. P., Wolin, M. J., "Ethyl Alcohol and Formate Metabolism in Extracts of S. *organism* Isolated from *Methanobacillus omelianskii*," *Bacteriol. Proc.* (1970) 134.

(24) Fina, L. R., Sincher, H. J., DeCou, D. F., "Evidence for Production of Methane from Formic Acid by Direct Reduction," *Arch. Biochem. Biophys.* (1960) **91**, 159–162.

(25) Stadtman, T. C., Barker, H. A., "Studies on the Methane Fermentation. IX. The Origin of Methane in the Acetate and Methanol Fermentations by Methanosarcina," *J. Bacteriol.* (1951) **61**, 81–86.

(26) Pine, M. J., Barker, H. A., "Studies on the Methane Fermentation. 12. The Pathway of Hydrogen in the Acetate Fermentation," *J. Bacteriol.* (1956) **71**, 644–648.

(27) Pine, M. J., Vishniac, W., "The Methane Fermentations of Acetate and Methanol," *J. Bacteriol.* (1957) **73**, 736–742.

(28) Paynter, M. J. B., Hungate, R. E., "Characterization of *Methanobacterium mobilis*, sp.n., Isolated from the Bovine Rumen," *J. Bacteriol.* (1968) **95**, 1943–1951.

(29) Langenberg, K. R., Bryant, M. P., Wolfe, R. S., "Hydrogen Oxidizing Methane Bacteria. 2. Electron Microscopy," *J. Bacteriol.* (1968) **95**, 1124–1129.

(30) Beijer, H. W., "Methane Fermentation in the Rumen of Cattle," *Nature* (1952) **170**, 576–577.

(31) Hungate, R. E., "The Rumen and Its Microbes," Academic, New York, 1966.

(32) Blaylock, B. A., "Cobamide-Dependent Methanol-Cyanocob(I)alamin Methyl Transferase of *Methanosarcina barkeri,*" *Arch. Biochem. Biophys.* (1968) **124**, 314–324.
(33) Wood, J. M., Allam, A. M., Brill, W. J., Wolfe, R. S., "Formation of Methane from Serine by Cell-Free Extracts of *Methanobacillus omelianskii,*" *J. Biol. Chem.* (1965) **240**, 4564–4569.
(34) Bryant, M. P., McBride, B. C., Wolfe, R. S., "Hydrogen Oxidizing Methane Bacteria. 1. Cultivation and Methanogenesis," *J. Bacteriol.* (1968) 1118–1123.

RECEIVED February 8, 1971.

Biochemistry of Methane Formation

B. C. McBRIDE and R. S. WOLFE

Department of Microbiology, University of Illinois, Urbana, Ill. 61801

Technology for the mass culture of methane bacteria on hydrogen–carbon dioxide gas mixtures has been developed to the extent that kilogram quantities of hydrogen-grown Methanobacterium *are now available. Carbon dioxide is reduced readily to methane by extracts, but the mechanism of carbon dioxide activation and reduction is unknown. Studies of methyl transfer reactions have been most fruitful with* Methanosarcina *and* Methanobacterium *where methylcobalamin is the substrate of choice. Methane formation requires ATP, but the nature of this reaction or the mechanism of ATP synthesis in methane bacteria is unknown. Active methyl groups may be transferred to arsenate with the formation of dimethylarsine or to mercury with the formation of methylmercury; these facts may be pertinent to potential dangers in anaerobic waste treatment.*

Eight years have elapsed since the preparation of the first active cell-free extracts of methane bacteria. With this event it was possible to examine the biochemistry of methane formation. At present only two laboratories are pursuing the biochemistry of methanogenesis by studying cell extracts of methane bacteria. This paper covers mainly work done at the University of Illinois.

When we started these investigations with *Methanobacillus omelianskii*, the culture was mass cultured in the ethanol–carbonate medium of Barker (*1, 2*), where ethanol was believed to be oxidized to acetate with the resulting electrons being transferred to reduce carbon dioxide to methane. About 1 gram of wet cells per liter of medium was obtained.

In collaborative studies with M. P. Bryant, this culture was found to be a symbiotic association of two organisms, one organism (S organism) oxidizing ethanol to acetate and hydrogen, the other organism (*Methano-*

bacterium) oxidizing hydrogen and reducing CO_2 to methane (3). This example of substrate coupling is shown in Reactions 1 and 2.

$$2H_2O + 2CH_3CH_2OH \xrightarrow{\text{S organism}} 2CH_3COOH + 4H_2 \qquad (1)$$

$$4H_2 + CO_2 \xrightarrow{\text{Methanobacterium (MOH)}} CH_4 + 2H_2O \qquad (2)$$

S organism is inhibited by the hydrogen it produces; *Methanobacterium* displaces the equilibrium by oxidizing hydrogen with reduction of carbon dioxide to methane. To survive in the ethanol–carbonate medium each organism requires the other.

With the discovery that hydrogen was the substrate for the methanogenic organism, which we call *Methanobacterium* strain MOH, it became necessary to evolve a technology for the mass culture of this strict anaerobe on hydrogen and carbon dioxide. A 200-ml growth flask was closed with a rubber stopper through which a mixture of oxygen-free H_2:CO_2 (80:20) was passed aseptically into the flask as it was incubated on a rotary shaker (4). The gas mixture was scrubbed of free oxygen by

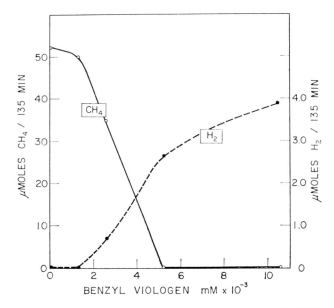

Journal of Bacteriology

Figure 1. Effect of varying concentrations of benzyl viologen on CH_4 and H_2 production in the presence of ethanol and CO_2. Gas atmosphere, 80% N_2:20% CO_2 (6).

passage through a heated copper column before being metered into the flask. To scale up the culture, a 14-liter fermentor was inoculated from the culture flask, and a $H_2:CO_2$ mixture was scrubbed and passed into the fermentor; a yield of 55–60 grams of wet cells was obtained from one of these fermentors (4). Results of a careful study of the stoichiometry of hydrogen oxidation during the growth of cells in a fermentor have shown that instead of a theoretical balance of 4 moles of hydrogen consumed per 1 mole of methane formed, a ratio of 3.7 to 1 was detected; this indicates that a small amount of reducing potential is available from components of the medium (5).

Viologen dyes are potent inhibitors of methane formation in whole cells (6). Methane formation is inhibited by increasing concentrations of benzyl viologen, and if the culture of *M. omelianskii* is used, molecular hydrogen is evolved as methane formation is inhibited (Figure 1). To explain the effectiveness of such low concentration levels it is believed that these dyes inhibit a specific site in the electron transport system by an irreversible reaction.

To study enzymic reactions, cell extracts were prepared by Hughes press or sonic probe (7), and activity was assayed in a Warburg flask which was closed with a rubber serum stopper. A hydrogen atmosphere was added to the flask, and after isolation of the flask, the substrate was tipped into the main compartment of the flask. Samples of the gas atmosphere were removed with a hypodermic syringe and were injected into a gas chromatograph which contained a silica gel column (7).

Table I.   Relative Rates of Methane Formation from Various Substrates

Flask[a]	Substrate	Methane Formed μmoles/hr/mg protein	Relative Rate of Methane Formation
1	$CO_2:H_2$ (80:20)	1.82	1.00
2	CH_3-B_{12} (5 μmoles)	1.03	0.57
3	Serine (10 μmoles)	0.67	0.37
4	N^5-CH_3H_4 folate (5 μmoles)	0.48	0.26
5	CH_3-cobaloxime (4 μmoles)	0.28	0.15

[a] Each reaction flask contained: cell extract, 30 mg protein; ATP, 10 μmoles; and TES [*N*-tris(hydroxymethyl)methyl-2-aminoethane sulfonic acid], 200 μmoles at pH 7.0; substrate as indicated; total volume 1.2 ml. Flasks 2–5 contained a hydrogen atmosphere.

Substrates which are precursors of methane in cell extracts are shown in Table I. These results were obtained with extracts of hydrogen-grown *Methanobacterium* (MOH). In contrast to earlier experiments with *M. omelianskii*, extracts of hydrogen-grown cells show greatest activity with CO_2. When the relative ability of these compounds to serve as

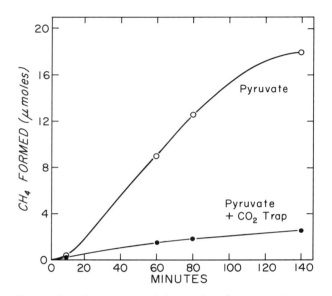

Figure 2. *Conversion of the carboxyl group of pyruvate to methane: effect of a CO_2 trap. Each flask contained sodium pyruvate, 10 μmoles; ATP, 10 μmoles; potassium phosphate, 750 μmoles at pH 7.0; hydrogen atmosphere; cell extract, 51 mg protein. The center well contained 0.3 ml 20% KOH solution and a filter paper wick as a CO_2 trap; the control flask contained 0.3 ml of water and a filter paper wick.*

precursors of methane is examined, the following order is obtained: $CO_2 > CH_3-B_{12} >$ carbon 3 of serine $> N^5-CH_3H_4$ folate $> CH_3$-cobaloxime. Historically, the first substrate to yield methane in cell extracts was pyruvate. The carboxyl group of pyruvate serves readily as a precursor of methane. As shown in Figure 2 the rate of methane formation from pyruvate is inhibited markedly when an alkaline CO_2 trap is added to the reaction vessel, an indication that methane is being formed *via* CO_2 and that an active C_1 intermediate is not formed directly from decarboxylation of pyruvate. Formate and methanol are not active in extracts of *Methanobacterium* (MOH). Extracts of *Methanobacterium formicicum* utilized formate in addition to the above compounds (4). Blaylock and Stadtman found that most of these compounds are active in *Methanosarcina,* and they were the first to extend the work of D. D. Woods with methylcobalamin to the methane bacteria (8). They found methylcobalamin to be an excellent substrate for studying transmethylation reactions. As shown in Figure 3 ATP is required to convert the methyl group of methylcobalamin to methane by extracts of *Methanobacterium* (9). However, high concentrations of ADP and AMP are inhibitory (Figure 4) (10).

We were interested in the specificity of methyl corrinoid derivatives and tested methyl factor III and methyl factor B as substrates for methane formation. Specificity for activation of the methyl group did not reside in the lower axial ligand since the methyl group was converted readily to methane in the absence of the dimethylbenzimidazole moiety of the lower axial ligand (*11*).

The work of Blaylock and Stadtman has significantly advanced our knowledge of the conversion of methanol to methane in *Methanosarcina*. A system has been developed in which B_{12s} serves as the methyl acceptor in the enzymic activation of methanol. These studies have revealed an unexpected complexity of this methyl transfer reaction; the requirements include ferredoxin, a corrinoid protein, an unidentified protein, ATP, Mg, a hydrogen atmosphere, and a heat-stable cofactor for the transfer of the methyl group of methanol to B_{12s} (*12*).

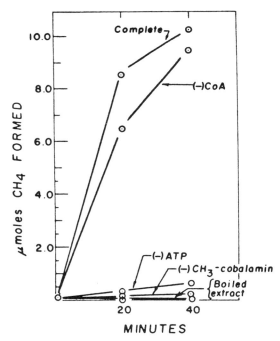

Biochemical and Biophysical Resaerch Communications

Figure 3. Effect of ATP on the formation of methane from methylcobalamin. The complete system contained crude extract, 45 mg protein; potassium phosphate buffer at pH 7.0, 500 μmoles; CoA, 0.05 μmole; ATP, 10 μmoles; methylcobalamin, 10 μmoles; total liquid volume 1.3 ml; hydrogen atmosphere; flasks incubated at 37°C. Values indicate total methane detected per flask (9).

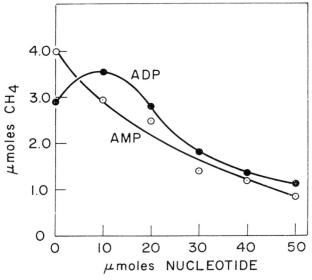

Figure 4. Effect of increasing concentrations of AMP or ADP on the formation of CH₄ in the presence of ATP. Reaction mixture contained crude extract, 41.0 mg of protein; methylcobalamin, 5.0 μmoles; ouabain, 50 μmoles; potassium phosphate buffer at pH 7.0, 760 μmoles; ATP, 10.0 μmoles; and the appropriate concentration of ADP or AMP. Total liquid volume, 2.0 ml. Reaction time, 30 min (10).

In *Methanobacterium* the cobamide product formed by removal of the methyl group from the cobalt atom is B_{12r}, a brown-colored cobamide derivative with cobalt in the 2+ valence state (*13*). We know that at least two protein fractions are required to carry out this reaction. Although a corrinoid protein (with 5-hydroxybenzimidazolylcobamide as the corrin moiety) was isolated readily from extracts of the culture known as *Methanobacterium omelianskii* and was shown to stimulate methane formation (*14*), this protein has not been isolated from extracts of hydrogen-grown *Methanobacterium*. To isolate this protein it was reduced in crude extracts, and the cobamide moiety was labelled with C-14 propyl iodide in the manner of Brot and Weissbach (*15*). Radioactivity was followed during fractionation of the protein. As shown in Figure 5, from a final Sephadex G-200 column, protein, radioactivity, and specific activity for stimulation of methane formation eluted as a single peak. To test this protein for enzyme activity in the methane-forming system the propyl moiety was removed by photolysis, and the protein was added to the partially resolved reaction mixture (*14*).

The observation of Bauchop that small amounts of chloroform inhibit methane formation in rumen fluid (*16*), prompted us to examine the effect of chlorinated hydrocarbons on methane formation in bacterial extracts. Chloroform, carbon tetrachloride, and methylene chloride were found to be competitive inhibitors of methane formation (*17*).

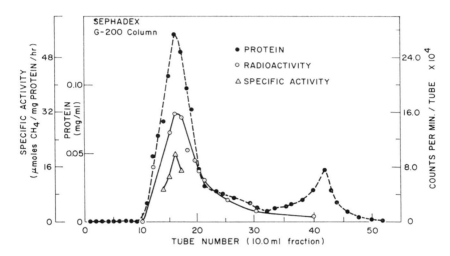

Biochemistry

Figure 5.   Comparison of enzymic activity (stimulation of methane formation) and radioactive specific activities of the [1-^{14}C] propyl-B_{12} enzyme from gel filtration on Sephadex G-200 (14).

The chloromethylcobalamin derivative of each halogenated hydrocarbon compound was synthesized by the reaction of the chlorinated hydrocarbon with B_{12s}. The similarity of the absorption spectra of these derivatives to that of methylcobalamin is shown in Figure 6. Photolysis breaks the cobalt–carbon bond, releasing methyl chloride. It was postulated that in the enzymic system the chlorinated hydrocarbon reacts with the cobamide protein to inhibit methane formation. To support this proposal it has been shown that inhibition of methane formation by these compounds is reversed by light. Studies with the pure cobamide enzyme from *E. coli*, 5-methyltetrahydrofolate homocysteine transmethylase (supplied by Taylor and Weissbach) have revealed a similar pattern of competitive inhibition upon photolysis (*17*). However, failure to detect the cobamide protein in hydrogen-grown cells indicates a significant change in the system which is not understood at present.

The role of ATP in methane formation also is poorly understood. When we first discovered the ATP requirement and tried to unravel the stoichiometry, the system required between 3 and 6 μmoles of ATP per

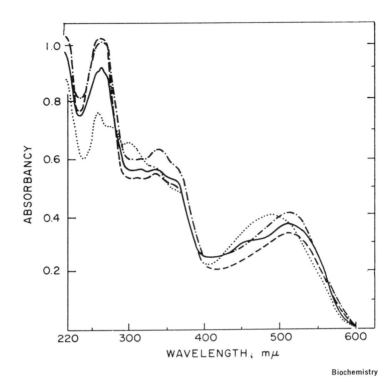

Biochemistry

*Figure 6.   Comparison of the absorption spectra of the chloro-
methylcobalamins with that of methylcobalamin in 0.1N acetic
acid.   (· · · · ·) Methylcobalamin, (———) chloromethylcobalamin,
(— · — ·) dichloromethylcobalamin, and (— — —) trichlorometh-
ylcobalamin (17).*

μmole of methane formed. Interpretations of the results were difficult
owing to the ATPase in the crude extracts. When high levels of ouabain
(a plant alkaloid which inhibits ATPase) were added; the ratio of ATP
hydrolyzed to methane formed approached a 1:1 ratio after substraction
of a significant blank. Hence, we proposed a scheme in which one ATP
would be utilized to make the coenzyme form of the cobamide moiety
which when attached to the protein could then accept a methyl group
from a variety of methyl donors (18). A reductive demethylation step
would form methane and regenerate the cobamide protein. This, of
course, would be a very expensive system for the cell to operate. We
now know that it is probably incorrect.

   A careful study of ATP metabolism in hydrogen-grown *Methanobacterium* by Roberton (19) has shown that the ATP requirement is not
stoichiometric. Less than one molecule of ATP is broken down per
molecule of methane produced; resynthesis of ATP is too slow to account

for this lack of stoichiometry. The results of an experiment where a hexokinase trap was added at a scheduled time to each of a series of identical reaction mixtures indicate that once ATP has reacted, free ATP per se is not an obligatory participant in methane synthesis from CO_2 and H_2.

It is not known how ATP is formed in methane bacteria; an ATP-yielding reaction has not been documented in cell extracts. A recent effort by Roberton to examine ATP pools in hydrogen-grown *Methanobacterium* has shown that energy conversion is very inefficient (5).

Methane bacteria have been shown to catalyze reactions in which the active methyl group is transferred to acceptors such as arsenate or mercury. When extracts are incubated in a hydrogen atmosphere with methylcobalamin, arsenate, and ATP, a volatile arsine derivative is formed (20). Arsines are difficult and dangerous to work with; they are extremely poisonous and are oxidized rapidly in air. Fortunately they have an intense garlic odor so the investigator is warned of their presence.

In reaction mixtures which contained arsenate the reaction mixture turned brown, indicating the transfer of the methyl group from methylcobalamin. Methane formation was inhibited. Of the various methyl donors tested as substrates only methylcobalamin could form an alkyl arsine; of the arsenic derivatives tested as substrates only cacodylic acid was reduced directly to an alkyl arsine without the addition of methylcobalamin. However, ATP and a hydrogen atmosphere were required, and the final alkyl arsine derivative was identified as dimethylarsine. The reductive pathway is as follows:

$$\text{Arsenate} \xrightarrow{2e} \text{Arsenite} \tag{3}$$

$$\text{Arsenite} \xrightarrow{[CH_3]} \text{Methylarsonic acid} \tag{4}$$

$$\text{Methylarsonic acid} \xrightarrow{[CH_3]\,2e} \text{Cacodylic acid} \tag{5}$$

$$\text{Cacodylic acid} \xrightarrow{4e} \text{Dimethylarsine} \tag{6}$$

This pathway may have considerable pertinence in anaerobic waste treatment when significant amounts of arsenate are present.

Selenate and tellurate also are potent inhibitors of methane formation, and we postulate that these compounds are reduced and methylated in a manner similar to that of arsenate.

Extracts of *Methanobacterium* transfer an active methyl group to mercury (21). Inhibition of methane formation from methylcobalamin upon the addition of mercury is presented in Figure 7. From such a reaction mixture small amounts of methylmercury and dimethylmercury

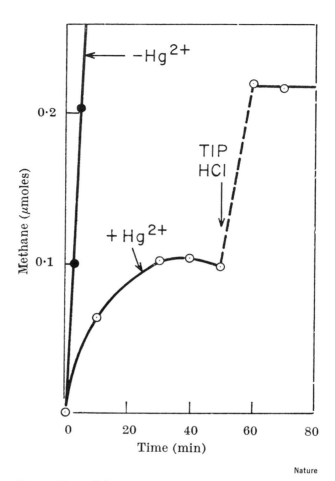

*Figure 7.   Inhibition of methane formation by Hg^{2+}
and the liberation of methane after the addition of hy-
drochloric acid.   Reaction contained: crude extracts,
56.2 mg of protein; 10.0 μmoles of ATP; 500 μmoles
of potassium phosphate buffer, pH 7.0; 0.1 μmole of
Hg^{2+} (where indicated), and 5.0 μmoles of methylcobalamin. The gas phase was H_2, the reaction volume
2.0 ml, and the incubation temperature 40°C. A sample
(0.5 ml) of 5N HCl was tipped into reaction mixtures
after 50 min (21).*

were detected. However, significant amounts of these compounds may
be formed chemically when methylcobalamin and mercury are incubated
under mild reducing conditions. The presence of mercury or derivatives
of mercury in anaerobic waste treatment digesters could lead to the
formation of highly toxic methylmercury compounds.

Schrauzer has synthesized a number of cobaloxime compounds which are considered analogs of cobalamin since the cobalt atom is in a similar position, being coordinated to four nitrogen atoms. Methylcobaloxime (with a methyl group as the upper axial ligand and water as the lower ligand) was tested as a methyl donor for CH_4 formation. Negative results were obtained until catalytic amounts of B_{12r} were added to the reaction mixture (*22*). As shown in Table II a number of methylcobaloxime derivatives in which the complexity of the lower ligand is varied have been tested. None of these compounds is as efficient as methylcobalamin in serving as a methyl donor for methane formation. These experiments have been assumed to demonstrate the biological activity of cobaloximes as analogs of vitamin B_{12}. However, until the requirement for B_{12r} in the reaction is understood, a cautious view of their biological significance should be taken.

Table II.   Rates of CH_4 Formation from Methylcobaloximes (*22*)

Substrate	Specific Activity of CH_4 Enzyme[a]
Methylcobalamin	15.20
Methyl-Co-(aquo)bis(dimethylglyoxime)	10.60
Methyl-Co-(aquo)bis(diphenylglyoxime)	9.60
Methyl-Co-(pyridine)bis(diphenylglyoxime)	9.50
Methyl-Co-(pyridine)bis(dimethylglyoxime)	8.40
Methyl-Co-(benzimidazole)bis(dimethylglyoxime)	3.50
Methyl-Co-(triphenylphosphine)bis(dimethylglyoxime)	3.40
Methyl-Co-(cyclohexyl isocyanide)bis(dimethylglyoxime)	1.30
Methyl-Co-(pyridine)bis(glyoxime)	1.20
Methyl-Co-(pyridine)bis(cyclohexanedione dioxime)	0.20

[a] Specific activity is defined as millimicromoles of CH_4 formed/mg protein/min. Each reaction contained 3.5 μmoles of methylcobaloxime derivative, 1.0 μmole of B_{12r}, 10 μmoles of ATP, and 100 μmoles of TES buffer, pH 7.0. Gas phase H_2, incubation temperature 40°C.

Progress toward an understanding of the biochemistry of methane formation has been slow because of the laborious nature of the mass culture of these strict anaerobes. In our laboratory the past year has been notable in that we can now obtain kilogram quantities of hydrogen-grown cells from a single fermentor. The time is ripe for a serious look at the unknown area of CO_2 activation and reduction.

It has become clear recently that a major substrate of methane bacteria in nature is hydrogen; all of the methane bacteria in pure culture in 1970 were able to use hydrogen. Many anaerobes which may be found in anaerobic waste treatment facilities produce hydrogen as a normal product of their metabolism, yet the amount of hydrogen detectable in sludge digesters is very low, an indication that it may be an important intermediate in methane formation.

Acknowledgment

It is a pleasure to acknowledge collaborations with Meyer J. and Eileen Wolin, Marvin Bryant, J. M. Wood, A. M. Robertson, and B. C. McBride in these experiments.

Literature Cited

(1) Barker, H. A., *J. Biol. Chem.* (1941) **137**, 153.
(2) Barker, H. A., "Bacterial Fermentations," Chap. 1, Wiley, New York, 1956.
(3) Bryant, M. P., Wolin, E. A., Wolin, M. J., Wolfe, R. S., *Arch. Mikrobiol.* (1967) **59**, 20.
(4) Bryant, M. P., McBride, B. C., Wolfe, R. S., *J. Bacteriol.* (1968) **95**, 1118.
(5) Roberton, A. M., Wolfe, R. S., *J. Bacteriol.* (1970) **102**, 43.
(6) Wolin, E. A., Wolfe, R. S., Wolin, M. J., *J. Bacteriol.* (1964) **87**, 993.
(7) Wolin, E. A., Wolin, M. J., Wolfe, R. S., *J. Biol. Chem.* (1963) **238**, 2882.
(8) Blaylock, B. A., Stadtman, T. C., *Biochem. Biophys. Res. Comm.* (1963) **11**, 34.
(9) Wolin, M. J., Wolin, E. A., Wolfe, R. S., *Biochem. Biophys. Res. Comm.* (1963) **12**, 464.
(10) Wood, J. M., Wolfe, R. S., *J. Bacteriol.* (1966) **92**, 696.
(11) Wood, J. M., Wolin, M. J., Wolfe, R. S., *Biochemistry* (1966) **5**, 2381.
(12) Blaylock, B. A., *Arch. Biochem. Biophys.* (1968) **124**, 314.
(13) Wolin, M. J., Wolin, E. A., Wolfe, R. S., *Biochem. Biophys. Res. Comm.* (1964) **15**, 420.
(14) Wood, J. M., Wolfe, R. S., *Biochemistry* (1966) **5**, 3598.
(15) Brot, N., Weissbach, H., *J. Biol. Chem.* (1965) **240**, 2064.
(16) Bauchop, T., *J. Bacteriol.* (1967) **94**, 171.
(17) Wood, J. M., Kennedy, F. S., Wolfe, R. S., *Biochemistry* (1968) **7**, 1707.
(18) Wolfe, R. S., Wolin, E. A., Wolin, M. J., Allam, A. M., Wood, J. M., "Developments in Industrial Microbiology," Vol. 7, p. 162, A.I.B.S., Washington, D. C., 1966.
(19) Roberton, A. M., Wolfe, R. S., *Biochim. Biophys. Acta* (1969) **192**, 420.
(20) McBride, B. C., Wolfe, R. S., *Bacteriol. Proc.* **1969**, 130.
(21) Wood, J. M., Kennedy, F. S., Rosen, C. G., *Nature* (1968) **220**, 173.
(22) McBride, B. C., Wood, J. M., Silbert, J. W., Schrauzer, G. N., *J. Am. Chem. Soc.* (1968) **90**, 5276.

Received August 3, 1970.

Nutrient Requirements of Methanogenic Bacteria

M. P. BRYANT, S. F. TZENG, I. M. ROBINSON,[1] and A. E. JOYNER, Jr.[2]

Departments of Dairy Science and Microbiology, University of Illinois, Urbana, Ill. 61801

Studies on nutrient requirements of one rumen strain (M1) and one sludge strain (PS) of Methanobacterium ruminantium *and* Methanobacterium *strain MOH, isolated from* Methanobacillus omelianskii, *indicate that these bacteria, grown with H_2–CO_2 as the energy source, require NH_4^+ as the main source of cell nitrogen and have little ability to incorporate or metabolize organic nitrogen compounds such as amino acids or peptides.* M. ruminantium *requires acetate for growth and synthesizes about 60% of cellular C compounds utilizing acetate C when grown in complex media. Strain MOH can utilize CO_2 as the main C source but is stimulated by acetate. Strain M1 also requires 2-methyl-butyrate and an unidentified factor for growth. This and other studies indicate that some of the more numerous methanogenic bacteria depend strongly on other bacteria in the natural habitat which supply essential sources of nutrients in addition to energy sources.*

While relatively little information is available on species of bacteria functional in anaerobic ecosystems such as the digestors of sewage treatment facilities so that definitive nutritional studies can be carried out, most species of bacteria functional in the rumen have been described, and many nutritional features of many of these species have been determined (*1, 2*). These nutritional studies of axenic rumen species have yielded much qualitative information, important to the understanding of this ecosystem, that would be difficult or impossible to obtain and interpret *via* studies on the nutritional characteristics of the mixed population (*3*).

---

[1] Present address: National Animal Disease Laboratory, Ames, Iowa.
[2] Present address: Shell Development Co., P.O. Box 4842, Modesto, Calif. 93552.

Qualitative information on pathways and important extracellular intermediates in the rumen catabolism of organic matter have been obtained—*e.g.*, the importance of succinate as a major extracellular intermediate in propionate formation was first indicated by the pure culture studies of Sijpesteijn (*4, 5*). Where the fermentation products of pure cultures did not seem to fit with the metabolism shown by the mixture in the natural habitat—*e.g.*, many pure cultures of rumen bacteria produce ethanol, yet ethanol is usually not a final product in the rumen and is usually not catabolized therein—new concepts of interspecies hydrogen transport and the effect of hydrogen utilizing methanogenic bacteria on energy metabolism and electron transport in carbohydrate fermenting bacteria were visualized (*2, 6*). Much definitive information on biochemical interactions between species has been obtained *via* nutritional studies. Previously unknown growth factors have been identified, and studies on their functions have led to discovery of important cellular constituents of bacteria—*e.g.*, plasmalogens in lipids (*7*)—and pathways of biosynthesis of cellular constituents—*e.g.*, reductive carboxylation of certain organic acids in the formation of amino acids (*8*). Addition of growth factors essential to certain microbial species (*e.g.*, branched-chain volatile acids) to certain ruminant diets stimulates microbial protein synthesis and feed conversion in the rumen (*9*). Knowledge of the nutrition of species selected by the rumen environment has given good information on the chemical characteristics of the environment. For example, finding that many rumen species cannot effectively utilize organic nitrogen compounds such as amino acids or peptides (*1, 10, 11*) helped lead to the concept that even though a ruminant diet may be relatively high in protein, there is relatively little organic nitrogen in the rumen environment which is available for growth of rumen bacteria (*12*).

Although few of the species of anaerobic bacteria functional in the more or less complete anaerobic dissimilation of organic matter to methane and carbon dioxide in systems such as anaerobic sewage digestors are known (*13*), we are beginning to recognize some of the species which are significant in the terminal stages—*i.e.*, the methanogenic species. Authentic pure cultures of many methane bacteria utilizing H_2–CO_2 and formate, and of at least one species, *Methanobacterium barkeri* which utilizes acetate, methanol, and CO, as energy sources for growth, are now available for study. Information on the nutrition of these methanogenic bacteria has been difficult to obtain because of problems in developing methods to obtain good growth yields in liquid media of known chemical composition from small inocula. This is caused primarily by their requirement for a very low oxidation-reduction potential for growth (*14, 15, 16, 17*). Also, they appear to be killed more rapidly by low levels of oxygen than are most strictly anaerobic bacteria (*16*).

Earlier studies on a few pure cultures, or highly purified cultures which contained few contaminants, established that methanogenic bacteria were restricted in energy sources utilized for growth and suggested that their nutritional requirements were quite simple (*18*). The species studied could be grown in defined media with ammonia as the nitrogen source and sulfide as the sulfur source. B-vitamins or other organic growth factors were not required, and complex materials, such as yeast extract, containing amino acids, nucleic acid degradation products, and vitamins were not stimulatory. Carbon and energy sources could be supplied by CO_2 and H_2 or a single organic compound such as methanol or formate. Among the pure cultures studied, *Methanobacterium formicicum* was grown with formate or CO_2 as the carbon source (*15, 19*). *Methanococcus vanniellii* also grew with formate as source of carbon (*20*) and *Methanosarcina barkeri* with methanol as the carbon source (*19*). *Methanobacillus omelianskii*, the organism on which the most detailed nutritional studies were done, grew with ethanol and CO_2 as the carbon sources (*21*), and the ability of this bacterium to fix nitrogen gas was demonstrated (*22*). In more recent studies the latter species was shown to be a synergistic association of two organisms (*14*), a methanogenic species which utilizes H_2–CO_2 as the energy source and a nonmethanogenic rod which produces acetate and H_2 from ethanol but fails to utilize ethanol effectively unless the partial pressure of H_2 is kept low. Whether one or both of these species fix N_2 is not known.

While these earlier studies supplied some good information on the nutrition of methane bacteria, they were conducted mainly on species isolated from elective enrichments, and the ecological significance of the species was in doubt except for *M. formicicum* which was shown to be present in large numbers in sludge (*15*).

In 1958 Smith and Hungate (*17*) described *Methanobacterium ruminantium*, which was present in large numbers in rumen content, and although detailed nutritional studies were not done, they showed that it required unknown growth factors which were present in rumen fluid but not found in many other nutritious materials such as yeast extract or peptones. In more recent studies, we confirmed and extended the nutritional information on *M. ruminantium* (*23*), and others have shown that other species require organic growth factors different from those required as energy source (*16, 24*).

The remainder of this paper discusses the nutritional requirements of three strains of methane bacteria grown with H_2–CO_2 as energy source —i.e., *M. ruminantium* strains isolated from sludge and the rumen and *Methanobacterium* strain MOH isolated from *M. omelianskii* (*14*). The latter is believed to be related closely to *M. formicicum*, except that it does not ferment formate (*25*). *M. formicicum* and *M. ruminantium* utilize only hydrogen and carbon dioxide or formate as energy source and for methane formation, and both are among the more numerous methanogenic bacteria found in sludge.

Nutrition of Methanobacterium ruminantium

Rumen Strain. In studies on strain M1 of *M. ruminantium* isolated from bovine rumen contents, we confirmed the results of Smith and Hungate (17), showing that this species requires factors present in rumen fluid (23). It would not grow in medium containing H_2–CO_2, bicarbonate, sulfide, cysteine, yeast extract, trypticase, and minerals unless rumen fluid was also added. In further experiments (23) we assayed fractions of rumen fluid using the above medium and showed that growth factors could be separated into two components by acidification and ether extraction. One factor was not extractable with ether, and others were shown to be volatile acids.

Further studies showed that, in fact, two volatile fatty acids present in rumen fluid were essential for growth of strain M1 (23). One of these was acetate, and relatively large amounts were required for optimal growth. In the complex assay medium indicated above, but containing the residue of acid–ether extracted rumen fluid, the optimal concentration of acetate was about 16 to 20 mM. Equimolar additions of NaCl or NaHCO$_3$ to the medium would not replace acetate.

The high requirement suggested that acetate was a major source of cell carbon in *M. ruminantium*. Some results of experiments with ^{14}C-acetate are shown in Table I. About 60% of the cell carbon is derived

Table I.   Incorporation of Acetate-^{14}C into *Methanobacterium ruminantium* during Growth in a Complex Medium Containing Rumen Fluid, Trypticase, and Yeast Extract[a]

	Specific Activity	
	(dpm per mg C)	
Component in Growth Medium	*Exp. 1*	*Exp. 2*
Acetate, zero time	23,400	29,500
Acetate, 69 hours incubation	24,700	26,000
Cells, 69 hours incubation	14,800	18,500
% cell C for acetate [b]	61.4	66.9

[a] Cells were grown in 400 ml of medium containing 2% clarified rumen fluid, 0.2% each of Trypticase, yeast extract, sodium formate, and sodium acetate, 0.01% of 2-methylbutyric acid and H_2–CO_2, minerals, Na$_2$CO$_3$, cysteine, and sulfide (26). Cells produced during 69 hours incubation at 37°C represented 122–150 mg dry wt/400 ml medium and carbon content of 42.0–43.8%. Acetate was isolated using the silica gel chromatographic method of Ramsey, and radioactivity measurements were made using liquid scintilation counting techniques.
[b] % cell C from acetate C = 100 × cell C specific activity/average specific activity of acetate C.

from acetate even when the organism is grown in a complex medium containing rumen fluid, Trypticase, yeast extract, formate, and CO_2 as other possible sources of carbon. Determination of ^{14}C in cell fractions obtained from sonically broken preparations using the fractionation procedures of Roberts *et al.* (*27*) showed that about 60, 20, and 14% of the acetate carbon was incorporated into the protein, nucleic acid, and lipid fractions, respectively. These studies indicate that acetate is a major source of cell carbon in *M. ruminantium* even when a complex variety of possible carbon sources are present in the growth medium.

The other volatile acid which is essential for growth of strain M1 is 2-methylbutyric acid (*23*). It is required in relatively small amounts in that a 0.05 mM concentration allows somewhat more than one-half maximal growth. Experiments by Robinson and Allison (*28*) show that more than 90% of ^{14}C-2-methylbutyrate incorporated by strain. M1 is incorporated into protein, and all this is present in isoleucine. These results indicate that *M. ruminantium* biosynthesizes isoleucine *via* the reductive carboxylation reaction as follows (*8*):

$$CH_3CH_2\overset{\overset{\displaystyle CH_3}{|}}{C}HCOOH + CO_2 + 4H + NH_3 \rightarrow CH_3CH_2\overset{\overset{\displaystyle CH_3}{|}}{C}H\underset{\underset{\displaystyle NH_2}{|}}{C}HCOOH + 2H_2O$$

It is suggested that *M. ruminantium* requires 2-methylbutyrate because it lacks the ability to assimiliate efficiently isoleucine from the medium or to biosynthesize the carbon skeleton of isoleucine from other carbon sources.

The growth factor for strain M1 associated with the ether extract residue of rumen fluid (*23, 29*) has been studied further, but it has not yet been identified. It is a relatively stable organic compound which is present in rumen fluid, effluent of anaerobic sewage digestors and among the cellular constituents of other methanogenic bacteria which do not require an exogenous source. For example, it is produced by strain PS, a sewage strain of *M. ruminantium,* a strain of *M. barkeri* and by *Methanobacterium* strain MOH. It is not produced by *E. coli* or a number of species of carbohydrate fermenting rumen bacteria, nor is it present in liver, peptones, or yeast extract. The amount of the factor present in rumen fluid can be more than doubled by *in vitro* incubation of fresh rumen contents for 24–48 hours with 0.4% of yeast extract added; however, additions of B-vitamins, nucleic acid, protein, starch, or cellobiose do not increase the amount. These results suggest that yeast extract contains a bound form or precursor which is converted to the active factor by microbial action.

The factor has been obtained in a highly purified form from rumen fluid by the following method. It is passed through a Dowex 50 (H form) column by elution with water to remove positively charged contaminants. This acid eluate is extracted exhaustively with ethyl acetate to remove contaminating lipids, and after evaporation to remove ethyl acetate the factor is adsorbed onto norite. After washing the norite with hot water to remove impurities the factor is eluted with $0.1M$ ethanolic NH_4OH and evaporated to small volume. At this stage the factor is still contaminated with a large amount of brown pigmented, and ultraviolet-light absorbing material but can be separated from these by partition chromatography on silicic acid with $0.05N$ H_2SO_4 as the stationary phase. After washing the column with about eight column volumes of 24% *tert*-butyl alcohol in chloroform, much of the factor elutes with 30% *tert*-butyl alcohol in chloroform. After two passages through this column, the factor is relatively pure as indicated by a lack of ultraviolet absorbing or visible pigment. Part of the factor remains on the column and can be removed only by stripping the column with higher concentrations of alcohol. This fact plus other data showing that two peaks of activity are found during gradient elution from TEAE cellulose column with NH_4HCO_3 suggest that two forms of the factor are present in rumen fluid. The information obtained indicate that the factor is a relatively low molecular weight, highly polar, relatively strong acid. We purified enough of the factor to obtain about half-maximal growth of a 25-liter culture of strain M1, but the dry weight of factor obtained was less than 1 mg. This indicates that it is required in a catalytic rather than in a substrate concentration. Pre-

Table II.   Effect of Amino Acids and Ammonia on Growth of Strain M1, *Methanobacterium ruminantium*

Additions to Basal Medium [a]	Growth OD $\times$ 100 (600 nm)
Experiment 1	
None	2
NH_4^+ (4 mM)	5
NH_4^+ (4 mM) + casamino acids (0.2%)	60
Experiment 2	
Amino acids (11 mM N)	4
Amino acids (44 mM N)	5
Amino acids (44 mM N) + NH_4^+ (4 mM)	59

[a] Basal medium contained 0.2% sodium formate, 5% mineral solution (18 grams KH_2PO_4, 18 grams NaCl, 0.53 grams $CaCl_2 \cdot 2H_2O$, 0.4 grams $MgCl_2 \cdot 6H_2O$, 0.2 grams $MnCl_2 \cdot 4H_2O$ and 0.02 gram $CoCl_2 \cdot 6H_2O$ in 1 liter), 0.3% volatile acid mixture (36 ml acetic, 14.8 ml propionic, 10.6 ml butyric, 1.8 ml isobutyric, and 2.0 ml each of *n*-valeric, isovaleric and DL-2-methylbutyric acids), 0.5% vitamin solution (20 mg each of thiamin·HCl, Ca-D-pantothenate, nicotinamide, riboflavin, and pyridoxine·HCl, 0.5 mg biotin, 0.25 mg folic acid, and 0.2 mg vitamin B_{12} in 100 ml water), 0.5% hemin solution (10 mg in 100 ml of 50% ethanolic, $0.1M$ KOH), and cysteine–sulfide, 1:1 H_2–CO_2 gas phase and Na_2CO_3 as previously indicated (14, 23). Rumen fluid (10%) was added after clarification by centrifugation, passage through a column of Dowex 50, H form, and neutralization to pH 6.5 with NaOH. Final pH was 6.7. Inoculum was as previously indicated (23) and growth measurements reported are mean values from three or four tubes. NH_4^+ was added as $(NH_4)_2SO_4$, and the amino acids added were a mixture of L-amino acids with kinds and proportions similar to those in casein.

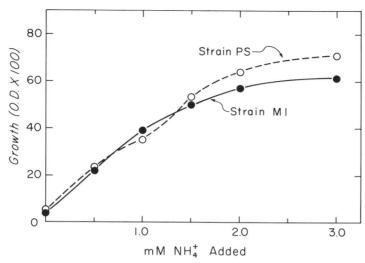

Figure 1.   Growth response of strain M1, a rumen strain of Methanobacterium ruminantium, *and strain PS, a sludge strain of* M. ruminantium, *to ammonia. The basal medium for strain M1 was the basal medium (Table II) plus 1.33 mM glycine, 0.95 mM L-arginine, 1.34 mM L-methionine, 0.48 mM L-histidine, and 1.68 mM L-threonine. The basal medium for strain PS was that indicated in Table II but with Dowex 50-treated rumen fluid and volatile acids other than acetate deleted. Ammonia was added as* $(NH_4)_2SO_4$, *and other conditions were as indicated in Table II.*

liminary studies suggest that this growth factor is identical with the dialyzable methyl-transfer coenzyme required for methane formation from CO_2 or methylcobalamin in extracts of *Methanobacterium* strain MOH (*30*). B. C. McBride, in the laboratory of R. S. Wolfe, is purifying this factor from large quantities of cells of strain MOH.

Even though strain M1 cannot be grown in defined media because of its requirement for the unidentified factor, some of its nitrogen requirements could be studied in media to which rumen fluid, treated with Dowex 50, H form, to remove nitrogen sources such as amino acids and ammonia, was added. Results in Table II show that both amino acids and NH_4^+ were essential to growth. Further studies on specific amino acids essential for growth were inconclusive, but good growth was never obtained in media containing only NH_4^+ and cysteine as possible nitrogen sources. Fairly good growth was obtained in media containing four to six amino acids, the most stimulating being threonine, histidine, and methionine.

Studies on the NH_4^+ requirement indicate that a large amount is required for growth (Figure 1). One μmole of NH_4^+ per ml of medium supplied enough nitrogen for growth to an optical density (OD) of about 0.36, and other experiments (Figure 2) showed that NH_4^+ utilized

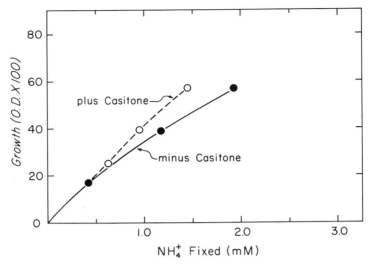

Figure 2.   Ammonia utilized by strain M1, Methanobacterium ruminantium, *in basal medium (Table II) with 2 mM (NH₄)₂SO₄, 1.33 mM glycine, 0.95 mM L-arginine, 1.34 mM L-methionine, 0.48 mM L-histidine, and 1.68 mM L-threonine added and in the same medium with Casitone (Bacto) added—about 30 mM Casitone nitrogen. NH_4^+ utilization was calculated as the difference in NH_4^+ present in the medium shortly after inoculation and after incubation to the given amount of growth. Ammonia was determined by microdiffusion and subsequent Nesslerization (10).*

during growth was similar whether or not a large amount of Casitone,— *i.e.,* amino acid and peptide nitrogen—was included in the medium. Analyses of total nitrogen (*10*) of washed cells indicated that a culture with growth equal to OD of 0.34 contained 1 μmole of cell nitrogen per ml. These experiments including those in Table II indicate that strain M1 utilizes ammonia as the main source of cell nitrogen whether or not the culture medium contains complex mixtures of amino acids or peptides.

Sewage Sludge Strain. The sludge strain PS of *M. ruminantium* is similar nutritionally to the rumen strain but is less exacting in its requirements and can be grown easily in a chemically defined medium—*i.e.,* the medium shown in Table II but with 3.8 m*M* (NH₄)₂SO₄ added and with rumen fluid and volatile acids other than acetate deleted. This strain does not require 2-methylbutyric acid, amino acids, or the unidentified factor required by the rumen strain. In fact, it biosynthesizes this growth factor. However, it does require acetate and ammonia.

Results in Figure 3 indicate that addition of materials such as casamino acids and yeast extract do not greatly affect growth in the defined medium. Growth in medium 1 was consistent and was maintained at about the same level through many serial transfers. The results show that one or more B-vitamins is essential for good growth of this strain. No work has been done on the identity of the B-vitamin required.

Table III shows that preformed exogenous amino acids or peptides are not utilized efficiently as nitrogen sources for strain PS. Whether or not Casitone was present in the medium, NH_4^+ utilized by growing cells was great. In fact, for a given amount of growth more NH_4^+ was utilized in the medium containing amino acids and peptides than in that contain-

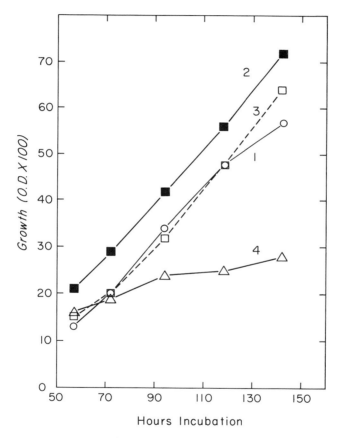

Figure 3. *Growth of strain PS in defined basal medium —i.e., that indicated in Table II but with 3.8 mM $(NH_4)_2SO_4$ added and with rumen fluid and volatile acids other than acetate deleted*

Curve 1: no additions
Curve 2: 0.4% Casamino acids (vitamin-free, Bacto) added
Curve 3: 0.4% Casamino acids and 0.4% yeast extract added
Curve 4: 0.4% Casamino acids added and vitamin mixture (Table II) deleted
Data indicate means of three tubes

Table III.   Ammonia Utilized by Strain PS of *Methanobacterium ruminantium* Grown in Defined Basal Medium with 2 mM $(NH_4)_2SO_4$ or with 2 mM $(NH_4)_2SO_4$ and 0.32% Casitone Added[a]

Tube No.	Growth $(OD \times 100)$	NH_4^+ Utilized (μmoles per Ml)	Growth per μmole NH_4^+ Utilized per Ml
	NH_4^+ as N Source		
1	19	0.512	37.1
2	33	1.113	29.7
3	36	1.163	31.0
mean	29	0.929	32.6
	NH_4^+ and Casitone as N Source		
1	14	0.543	25.8
2	17	0.593	28.7
3	36	1.418	25.4
4	36	1.408	25.6
5	47	1.768	26.6
mean	30	1.146	26.4

[a] Basal medium was identical to medium 1, Figure 3 but with $(NH_4)_2SO_4$ added as indicated. NH_4^+ utilization was determined as indicated in Figure 2. The Casitone added to the medium represented about 30 mM nitrogen in amino acids plus peptides.

ing only NH_4^+ as the nitrogen source. Comparison of results in Table III on NH_4^+ uptake for a given amount of growth with those in Figure 1 on growth response to a given amount of NH_4^+ when it was the sole source of nitrogen in the medium also suggests that most of the cell nitrogen is derived from ammonia even when a complex source of organic nitrogen is present in the medium. Results in Figure 4 show that the sludge strain, like the rumen strain of *M. ruminantium,* requires a relatively large amount of acetate for growth.

Interesting observations concerning the sulfur requirements of strain PS were made during the development of methods for producing large amounts of cells for biochemical studies. The general methods used were as reported previously (26) except that the MF 128, steam-sterilizable, Microferm (New Brunswick Scientific Co.) with 20-liter batches of medium (rather than 12-liter batches) was used. The growth medium contained 0.4% trypticase and 0.2% each of yeast extract, sodium acetate, and sodium formate, and the usual minerals including 0.075% $(NH_4)_2SO_4$, and 0.05% cysteine · HCl. After sterilization, 600 ml of sterile, CO_2-equilibrated, 8% Na_2CO_3 and 50 ml of sterile, N_2-equilibrated, 20% $Na_2S · 9H_2O$ were added. After inoculation with 400 ml of active culture, 1:1 mixture of H_2–CO_2 was sparged through the culture at a rate of 400 ml/minute with stirring at 275 rpm to maintain a fine dispersion of gas. In these cultures growth was initiated but soon stopped with yields of

0.2–0.4 gram wet weight/liter. Additions of further Na₂S resulted in more growth. Since phenosafranin remained reduced in these cultures, it was concluded that sulfide was essential as sulfur source rather than as a reducing agent, but it was lost rapidly as H₂S under these conditions of sparging and pH (*ca* 6.7). When the fermentor was modified so that a small amount of H₂S (about 1 ml/minute) was passed continuously through the fermentor with the H₂–CO₂ gas, yields were increased to 2–3 grams/liter. These results suggest strongly that sulfide is the source of sulfur for this organism and that cysteine, methionine, and sulfate do not serve as sulfur sources.

Nutrition of Methanobacterium *Strain MOH*

This organism utilizes only H₂–CO₂ as the energy source, and a detailed study of its nutrition (*31*) indicates that it is similar nutritionally to the sewage sludge strain of *M. ruminantium* but has greater biosyn-

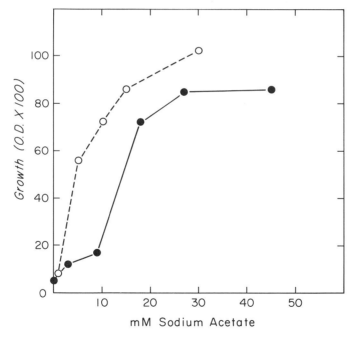

Figure 4.   Growth response of strain PS to sodium acetate in the defined basal medium—i.e., that indicated in Table II but with 3.8 mM (NH₄)₂SO₄ added and with rumen fluid and volatile acids other than acetate deleted. NaCl (○) or Na₂CO₃ (●) were added to maintain the normality of Na⁺ constant with the varying concentrations of acetate. Data indicate means of three tubes.

thetic abilities in that acetate is not required for growth. Results in
Figure 5 show that NH_4^+ serves as the main nitrogen source and that
cysteine, present in all of these media, does not serve as the nitrogen
source nor do amino acids or peptides present in Casitone. The small
amount of growth in media containing Casitone but not added NH_4^+ is
caused by the small amount of NH_4^+ that contaminates Casitone. The
results also show that acetate is quite stimulatory in the presence or
absence of Casitone but is not required for growth, suggesting that CO_2

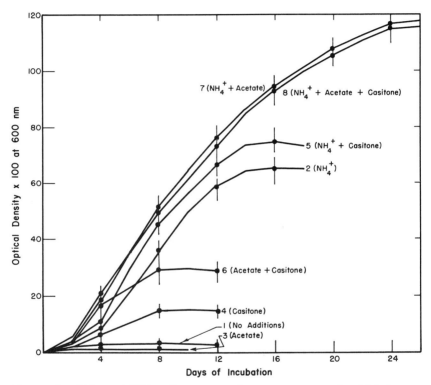

Figure 5. Effect of NH_4^+, Casitone, and acetate on growth of Methanobacterium *strain MOH. The basal medium contained 1.0% heavy metals
solution (32) and B-vitamins, minerals, Na_2CO_3, cysteine, and sulfide, and
1:1 H_2–CO_2 gas phase as indicated for the basal medium in Table II.*

Curve 1: no additions
Curve 2: 4 mM $(NH_4)_2SO_4$
Curve 3: 12.2 mM sodium acetate
Curve 4: 0.2% Casitone (vitamin-free, Bacto)
Curve 5: $(NH_4)_2SO_4$ and Casitone
Curve 6: Casitone and acetate
Curve 7: $(NH_4)_2SO_4$ and acetate
Curve 8: $(NH_4)_2SO_4$; acetate and Casitone
OD determination represent means of three tubes, and the vertical lines repre-
sent the range

can serve as the main source of carbon in this species. However, all media contained a small amount of cysteine (1.42 mM). Results, some of which are presented in Table IV, show that the amount of acetate required for optimal growth, about 5 mM, is less than that required by *M. ruminantium*.

Table IV. Effect of Acetate on Growth of *Methanobacterium* Strain MOH

Additions to Basal[a], mM		Growth (OD × 100)		Incubation Time, days
Na acetate	NaCl	mean[b]	range	
0	20	56	48– 62	12
1	19	76	70– 82	17
3	17	93	87– 96	18
5	15	105	95–110	21
10	10	110	105–120	23

[a] Basal medium was that shown in Figure 5 except 4.0 mM $(NH_4)_2SO_4$ was added.
[b] Mean of three replicate tubes at maximal OD.

To obtain further data to support the possibility that CO_2 can serve as the sole source of carbon other than that in B-vitamins, further experiments were performed. The organism could be maintained with difficulty in serial transfer in the basal medium (Figure 5) with $(NH_4)_2SO_4$ added as nitrogen source and with Na_2S added as sole reducing agent. However, growth was poor—i.e., the OD never was higher than about 0.3 compared with a control OD of about 0.85 when 1.42 mM cysteine was added. This was undoubtedly caused by the fact that the medium must be gassed each day to replenish H_2 energy source and, with the pH of the medium about 6.7, much of the sulfide is lost as H_2S. The medium tends to oxidize because of this and contamination with traces of air. Similar results were obtained in comparison of the reducing agents, cysteine and sulfide, in work with *M. ruminantium* (23). Dithiothreitol is an effective nonvolatile reducing agent which maintains the medium in a reduced state, but strain MOH failed to grow when it was the possible source of carbon other than CO_2 or source of sulfur other than sulfate in the medium (Figure 6). However, the results show that when sulfide or cysteine was added with dithiothreitol, growth was good. This indicates that sulfide or cysteine serves as sulfur source while sulfate or dithiothreitol do not; since the latter does not serve as the sulfur source, it is unlikely to serve as a significant source of carbon for the organism. Thus, it seems probable that CO_2 can serve as the main source of carbon for growth of strain MOH. Further studies on incorporation of $^{14}CO_2$ during growth in media with and without acetate should be of interest.

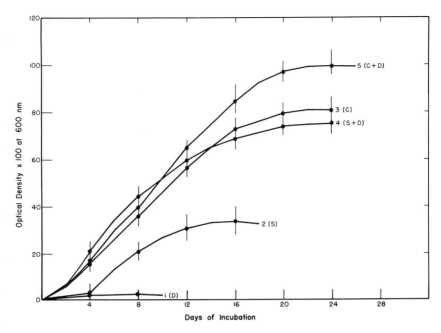

Figure 6. Effect of dithiothreitol, cysteine, and sulfide on growth of Methanobacterium *strain MOH in basal medium with CO_2 as carbon source. The basal medium was that shown in Figure 5 but with 4 mM $(NH_4)_2SO_4$ added and with cysteine and sulfide deleted. Additions were as follows:*

Curve 1: 2.52 mM dithiothreitol (D)
Curve 2: 2.10 mM Na₂S (S)
Curve 3: 2.84 mM cysteine (C)
Curve 4: S plus D
Curve 5: C plus D

As with *M. ruminantium, Methanobacterium* strain MOH can utilize neither peptides nor amino acids as a nitrogen source replacing NH_4^+ (Figures 5 and 7). NH_4^+ utilization is essentially the same (Figure 8) when NH_4^+ is the nitrogen source or when rumen fluid, amino acids, yeast extract, and Trypticase are present in the medium as complex sources of nitrogen that would be utilized in place of NH_4^+ by most heterotrophic bacteria other than some rumen carbohydrate-fermenting anaerobes (*11*).

Studies on B-vitamin requirements of strain MOH (*31*) were not completely definitive, but one or more is either very highly stimulatory or essential. The organism could be transferred serially in the basal medium (Figure 5) with B-vitamins deleted and with acetate and NH_4^+ added, but growth was poor (OD 0.25 or less). Deletion of biotin from the medium with B-vitamins added caused markedly depressed growth and, in some experiments, deletion of vitamin B_{12} or folic acid depressed growth significantly.

Discussion and Conclusions

Methane fermentation in natural ecosystems is usually described as a two-stage system in which nonmethanogenic bacteria ferment organic matter such as carbohydrate yielding such products as acetate, formate, hydrogen, and carbon dioxide. The methanogenic bacteria are restricted

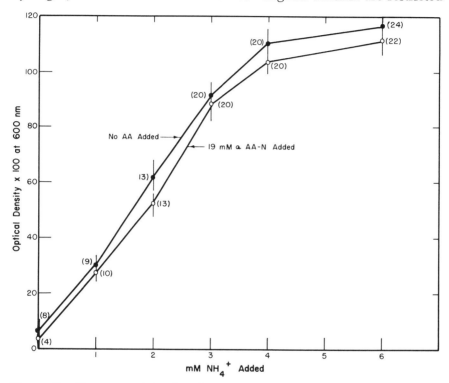

Figure 7. Response of Methanobacterium *strain MOH to growth-limiting concentrations of* NH_4^+ *in the presence and absence of free L-amino acids (19 mM α-amino acid nitrogen). The basal medium was that shown in Figure 5 but with 12.2 mM sodium acetate added. The proportions and kinds of amino acids were similar to those of casein. Numbers in parentheses indicate days of incubation for maximal OD.*

greatly in substrates utilized as energy sources and probably utilize only these fermentation products of other bacteria. The nutritional studies on hydrogen-utilizing methane bacteria reported here indicate that in nutritional requirements other than energy sources, these bacteria also depend on products produced by other bacteria. Thus, ammonia is produced from organic nitrogen compounds by other bacteria in the first stage of methane fermentation, and the methane bacteria studied so far require it as the main nitrogen source. The experiments reported indicate that amino acids, peptides, or other nitrogen compounds as present in crude materials are not utilized as major nitrogen sources. The requirement or stimulatory effect of acetate on these bacteria and the experiment indicating that acetate accounts for about 60% of the cell carbon in *M. ruminantium* grown in a complex medium support the conclusion

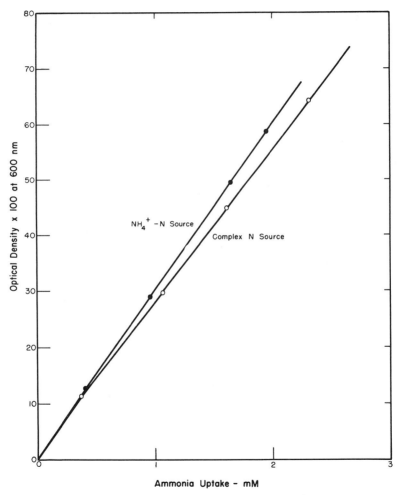

Figure 8. NH$_4^+$ utilized during growth of Methanobacterium *strain MOH in media with NH$_4^+$ as nitrogen source and with complex nitrogen sources added. Medium was the same as the basal shown in Figure 5 but with 12.2 mM sodium acetate and 2 mM (NH$_4$)$_2$SO$_4$ added or the same medium but with 20% of clarified rumen fluid, 0.5% of Trypticase, 0.2% of yeast extract and the mixture of L-amino acids added (Figure 7). Ammonia fixed was determined as indicated in Figure 2.*

that exogenous materials such as amino acids or peptides are not utilized effectively and show that at least some of these hydrogen-utilizing bacteria depend on other bacteria which produce acetate. 2-Methylbutyric acid is produced from isoleucine by amino acid-fermenting bacteria, and, thus, some strains of *M. ruminantium* depend on these bacteria. Although *Methanobacterium* strain MOH does not require 2-methylbutyrate for growth, Robinson and Allison (*28*) showed that *M. omelianskii* synthesizes isoleucine from 2-methylbutyrate, suggesting that strain MOH also utilizes 2-methylbutyrate.

The reason for the inefficiency of these bacteria in utilizing amino acids is not definitely known; however, it seems probable, as is the case with some of the carbohydrate-fermenting rumen bacteria (*8, 33*), that these materials are not transported effectively into the cell. It seems probable that, in their adaptation to the natural habitat, their abilities to utilize proteins and amino acids were of little survival value and were lost. It is well known that amino acids are present in very small amounts in extracellular fluid of rumen contents while relatively high concentrations of ammonia and acetate and other volatile acids including 2-methyl-butyrate are present (*8*). A similar condition probably exists in sludge although the volatile acids are usually present in lower concentration than in the rumen.

The present studies confirm the earlier studies indicating the relatively great biosynthetic abilities of the methane bacteria and suggest that much of the cellular carbon compounds are probably synthesized from acetate and carbon dioxide. In view of the carbon dioxide and acetate requirements and the reductive carboxylation reactions shown to be involved in isoleucine synthesis in *M. ruminantium* (*26*) and the probability of similar carboxylation reactions in biosynthesis of isoleucine, alanine, and other amino acids in MOH, suggested by the studies on *M. omelianskii* (*34*), the operation of the pyruvate synthase reaction and some other reactions of the reductive carboxylic acid cycle (*35, 36*) as major pathways of biosynthesis of cellular materials in these bacteria is an attractive hypothesis.

Acknowledgment

The authors acknowledge the excellent technical assistance of Melvin R. Crabill and Kenneth Holmer in various aspects of this work. Strain PS, *M. ruminantium*, was obtained through the courtesy of Paul H. Smith, University of Florida, Barry McBride kindly supplied hot 80% ethanol extracted-Dowex 50-treated extracts of strains MOH, PS and *Methanosarcina* for assay of unidentified factor required by strain M1. This research was supported in part by PHS research grant EC00289 and U.S. Department of Agriculture Hatch funds 35-331.

Literature Cited

(1) Bryant, M. P., Robinson, I. M., *J. Bacteriol.* (1962) **84**, 605.
(2) Hungate, R. E., "The Rumen and Its Microbes," Academic, New York, 1966.
(3) Allison, M. J., "Physiology of Digestion in the Ruminant," R. W. Dougherty, Ed., p. 369, Butterworth, Washington, D. C., 1965.
(4) Sijpesteijn, A. K., *J. Gen. Microbiol.* (1951) **5**, 869.
(5) Sijpesteijn, A. K., Elsden, S. R., *Biochem. J.* (1952) **52**, 41.
(6) Wolin, M. J., "Abstracts of Papers," 158th National Meeting, ACS, 1969, MICR 19.
(7) Allison, M. J., Bryant, M. P., Katz, I., Keeney, M., *J. Bacteriol.* (1962) **83**, 1084.
(8) Allison, M. J., *J. Animal Sci.* (1969) **29**, 797.
(9) Hemsley, J. A., Moir, R. J., *Austral. J. Agric. Res.* (1963) **14**, 509.
(10) Bryant, M. P., Robinson, I. M., *Appl. Microbiol.* (1961) **9**, 96.
(11) Bryant, M. P., Robinson, I. M., *J. Dairy Sci.* (1963) **46**, 150.
(12) Allison, M. J., "Physiology of Digestion and Metabolism in the Ruminant," A. T. Phillipson, Ed., p. 456, Oriel, Newcastle upon Tyne, England, 1970.
(13) Mah, R. A., Sussman, C., *Appl. Microbiol.* (1968) **16**, 358.
(14) Bryant, M. P., Wolin, E. A., Wolin, M. J., Wolfe, R. S., *Arch. Mikrobiol.* (1967) **59**, 20.
(15) Mylroie, R. L., Hungate, R. E., *Can. J. Microbiol.* (1954) **1**, 55.
(16) Paynter, M. J. B., Hungate, R. E., *J. Bacteriol.* (1968) **95**, 1943.
(17) Smith, P. H., Hungate, R. E., *J. Bacteriol.* (1958) **75**, 713.
(18) Barker, H. A., "Bacterial Fermentations," Wiley, New York, 1956.
(19) Schnellen, C. G. T. P., "Onderzoekingen Over de Methaangisting," Ph.D. Thesis, Technical University, Delft (1947).
(20) Stadtman, T. C., Barker, H. A., *J. Bacteriol.* (1951) **62**, 269.
(21) Barker, H. A., *Antonie v. Leeuwenhoek J. Microbiol. Serol.* (1940) **6**, 201.
(22) Pine, M. J., Barker, H. A., *J. Bacteriol.* (1954) **68**, 589.
(23) Bryant, M. P., "Physiology of Digestion in the Ruminant," R. W. Dougherty, Ed., p. 411, Butterworth, Washington, D. C., 1965.
(24) Smith, P. H., "Developments in Industrial Microbiology," Vol. 7, p. 156, American Institute of Biological Sciences, Washington, D. C., 1966.
(25) Langenberg, K. F., Bryant, M. P., Wolfe, R. S., *J. Bacteriol.* (1968) **95**, 1124.
(26) Bryant, M. P., McBride, B. C., Wolfe, R. S., *J. Bacteriol.* (1968) **95**, 1118.
(27) Roberts, R. B., Abelson, P. H., Cowie, D. B., Bolton, E. T., Britton, R. J., "Studies in Biosynthesis in *Escherichia coli*," *Carnegie Inst. Wash. Publ.* **607** (1957).
(28) Robinson, I. M., Allison, M. J., *J. Bacteriol.* (1969) **97**, 1220.
(29) Bryant, M. P., Nalbandov, O., *Bacteriol. Proc.* (1966) 90.
(30) McBride, B. C., Wolfe, R. S., *Fed. Proc.* (1970) **29** (2), 344.
(31) Tzeng, S. F., "Nutrition of *Methanobacterium* Strain MOH Isolated From *Methanobacillus omelianskii*," M.S. Thesis, University of Illinois (Urbana) (1970).
(32) Pfennig, N., Lippert, K. D., *Arch. Mikrobiol.* (1966) **55**, 258.
(33) Pittman, K. A., Laksmanan, S., Bryant, M. P., *J. Bacteriol.* (1967) **93**, 1499.
(34) Knight, M., Wolfe, R. S., Elsden, S. R., *Biochem. J.* (1966) **99**, 76.
(35) Buchanan, B. B., Evans, M. C. W., Arnon, D. I., *Arch. Mikrobiol.* (1967) **59**, 32.
(36) Evans, M. C. W., Buchanan, B. B., Arnon, D. I., *Proc. Natl. Acad. Sci.* (1966) **55**, 928.

RECEIVED August 3, 1970.

4

Volatile Acid Formation in Sludge Digestion

DAVID P. CHYNOWETH[1] and ROBERT A. MAH

Department of Environmental Sciences and Engineering, University of North Carolina, Chapel Hill, N. C. 27514

Addition of the normal raw sludge feed to a sample of digesting sludge from a domestic treatment plant results in an immediate increase in the rate of formation of gas and volatile acids, particularly acetic acid. The component in the raw sludge responsible for the increased acetogenic activity is mainly in the lipid fraction of the solids. Palmitic acid is fermented to acetate more readily than other pure substrates of the carbohydrate or protein class. Molecular hydrogen is not produced in these studies when methanogenic inhibitors such as $CHCl_3$ are added. The rapid dissimilation of lipids without a lag indicated existence of the appropriate conditions for growth of lipid-decomposing organisms.

Short chain organic acids, some neutral compounds, and H_2 and CO_2 are the main end products of the fermentation of organic compounds by pure cultures of bacteria. In the anaerobic sludge digestion process, such products serve as substrates for methane formation by the mixed bacterial population. If the over-all fermentation is vigorous, these methanogenic substrates are immediately converted to CH_4 and CO_2, and hence they do not accumulate, The net result is a continuous degradation of the complex starting substrates to gaseous end products, For convenience, the fermentation has been regarded as a two-stage process composed of an acid-forming phase and a methane-forming phase, each of which is carried out by two physiologically separate groups of bacteria. It should be recognized, however, that in the digester there is a continuum of metabolic reactions probably brought about by a complex intimate association between the so-called "acid formers" and the methane-producing bacteria.

---

[1] Present address: Department of Environmental and Industrial Health, School of Public Health, University of Michigan, Ann Arbor, Mich. 48104.

Acetate, which may be regarded as an intermediate fermentation product in this system, is responsible for over 70% (1, 2) of the total methane produced during the sludge fermentation. Despite this role, little information is available on the actual acetogenic substrates or the organisms responsible for their conversion to acetate in the unenriched system. This is partly the result of the fact that sludge is composed of a complex mixture of various carbohydrates, proteins, and lipids. The over-all fermentation depends consistently on acetate as a central intermediate even though the composition of the substrates may vary from one time to another. Our studies are concerned with the over-all fermentation of substrates by unenriched sludge, the role of non-methanogenic bacteria in these conversions and, in particular, the formation of acetate from various sludge fractions.

Table I. Composition of Raw Sludge

Reference	% Volatile Solids				% Total solids
	Carbo-hydrate	Proein	Lipid	Total	Volatile solids
Balmat (3)	30.2	32.3	24.5	87.0	78.0
Buswell and Neave (4)	17.7	31.8	41.4	90.9	60.9
Heukelekian (5)	16.9	35.7	45.3	97.9	75.9
Heukelekian and Balmat (5)	30.2	30.8	23.5	84.5	65.0
Hunter and Heukelekian (7)	43.5	19.4	18.4	80.9	81.0
Maki (8)	62.4	29.2	21.5	113.1	65.1
O'Rourke (9)	—	21.6	23.4	—	79.6
Average	33.5	28.5	28.2	90.1	72.2

Variation in the composition of domestic sewage does occur. Table I shows the values reported by several investigators (3, 4, 5, 6, 7, 8, 9) for the "volatile solids" (total solids minus ash) and for three major classes of organic compounds, carbohydrates, lipids, and proteins. All three occurred in roughly equal proportions; carbohydrates averaged 33.5%, and lipids and proteins each approximately 28% of the total organic compounds present. Actual volatile acids in the raw undigested sludge comprised only about 3–4% of the total organic solids. Although the concentration of these three components was reduced after digestion, it was surprising to note (Table II) that carbohydrates were decomposed only by about 13% (average of two values) and were least affected by the fermentation process. Proteins were decomposed about 36% (average of five values) and lipids about 76% (average of seven values). Thus, it might be suspected that lipids are the most important substrate in the sludge fermentation because of their high degree of degradability. The anaerobic dissimilation of long chain fatty acids leads to the formation

Table II.   Reduction of Substrates in Raw Sludge During Digestion

Reference	% Reduction			
	Carbo-hydrate	Protein	Lipid	Volatile solids
Buswell and Neave (*4*)	9.3	35.5	72.6	35.2
Buswell, Symons, and Pearson (*10*)	—	63.5	90.3	—
Heukelekian (*5*)	17.6	27.4	76.2	42.0
Heukelekian and Mueller (*11*)	—	—	75–89	—
O'Rourke (*9*)	—	17.0	65.2	—
Rudolfs (*12*)	—	—	70–80	—
Sawyer and Roy (*13*)	—	—	71	—
Woods and Malina (*14*)	—	38	—	—

of acetate as a primary breakdown product, probably by an anaerobic–oxidation reaction as indicated by the findings of Jeris and McCarty (*15*). There is, however, no direct evidence to show that lipids, particularly long chain fatty acids, are the chief acetogenic substrate in the normal sludge fermentation.

Previous workers (*1, 15, 16, 17, 18, 19*) examined the conversion of organic compounds to CH_4 and CO_2 after an initial period of enrichment of the sludge on the test substrate. The results of Mah *et al.* (*20*) showed that addition of excess concentrations of soluble carbohydrates such as glucose, cellobiose, or sucrose led to the enrichment of a euryoxic (*i.e.*, facultatively anaerobic) population of bacteria which bore no resemblance to the original population of anaerobes. This change occurred within 10 to 12 hours after addition of the test substrate. The euryoxic organisms comprised less than 2.4% of the starting population but were among the most numerous types after this comparatively short enrichment period. Similarly, "acclimatization" of laboratory digesters to other synthetic substrates must also lead to selection of a population of organisms different from those in the starting system. The organisms which best metabolize the test substrate will predominate. The studies reported in the present investigation were conducted on digesters unacclimatized to the particular substrate examined.

Materials and Methods

Sludge. The source of sludge for this study was a laboratory digester inoculated with a sample from the primary digester at the Third Fork Sewage Treatment Plant in Durham, N. C. The wastes treated at this plant are primarily of domestic origin. The digester was maintained on a raw sludge feed under conditions previously described (*2, 21*).

Anaerobic Methods. Strict anaerobic procedures were used throughout this investigation. Sludge samples were flushed with O_2-free 70% N_2–30% CO_2 (*2*) whenever handling was necessary.

Manometric Procedures. The rates of fermentation of various substrates were measured manometrically by the method of Smith and Mah (2). Special 100-ml Warburg vessels contained 25 ml of sludge.

Lipid Extration Procedures. Lipid was extracted from raw sludge with chloroform and methanol according to the method of Bligh and Dyer (22). Prior to use as a substrate, the lipid extract was dispersed in 2% Tween 80 by sonicating for 30 minutes at 60°C.

Volatile Acids and Ethanol. Volatile acids and ethanol were determined by gas chromatographic analysis using an Aerograph HiFi (model 550-B) equipped with a gold-plated flame ionization detector. Before analysis, samples were acidified with 3% metaphosphoric acid and centrifuged. The injection volume was 3 μliters. The acids and ethanol were separated at 135°C in a 9-ft Teflon column packed with Resoflex standard concentration P (Burrell Corp., Pittsburgh, Pa.). The column was packed by vibration and conditioned at 150°C. N_2 zero gas served as carrier gas at a flow rate of about 20 ml/minute. The H_2 flow rate was maintained by a pressure of 34 psi and the O_2 at about 400 ml/minute.

The Wiseman-Irvin method (23) of celite chromatography was used to separate the formate, acetate, propionate, and butyrate produced from ^{14}C-labelled substrates. Prior to chromatography, 2-ml samples were acidified with 2 ml cold carrier acids (0.04N each of formate, acetate, propionate, and butyrate) in 5N H_2SO_4.

Radioisotope Procedures. Radioactive volatile acids separated by Wiseman-Irvin chromatography were collected in 1 ml of 0.5N KOH. The aqueous fraction containing the acid was transferred to a scintillation vial and evaporated to dryness in a vacuum oven at 20 psi and 50°C. The residue was redissolved in 0.1 ml distilled H_2O before adding 4 ml of absolute ethanol and 15 ml of scintillation fluid: 2,5-diphenyloxazole (PPO) and 0.01% 1,4-bis[2-(5-phenyloxazolyl)]benzene (POPOP) in toluene. Samples were counted in a liquid scintillation counter (Nuclear Chicago Corp.).

The following U-^{14}C substrates were incubated with digesting sludge at 35°C for 60 minutes to determine the rates of formation of volatile acids: protein hydrolysate, pyruvate, glucose, glycerol, ethanol, and succinate. Lactate-1-^{14}C was also tested. At 10-minute intervals, samples were removed and killed by addition of an equal amount of 5N H_2SO_4 containing cold carrier acids as previously described. Volatile acids were then recovered by column chromatography, and the fractions containing radioactivity were determined by scintillation counter.

Inhibitor Studies. Test tubes containing 10 ml of digesting sludge and the appropriate substrates were incubated at 35°C with varying concentrations of CCl_4 and $CHCl_3$ and in later experiments with 3 μM $CHCl_3$. The gas phase was analyzed by gas chromatography (2), and the samples were killed with 4% metaphosphoric acid. Volatile acids were determined by gas chromatography as previously described.

Results

After the addition of the normal raw sludge substrate (Figure 1), the acetate concentration doubled immediately at zero time because of the presence of acetate in the substrate; thereafter, acetate continued to

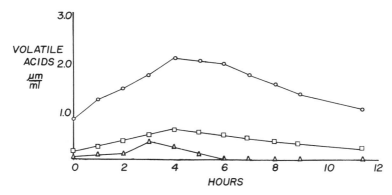

Figure 1.   Volatile acid production during the fermentation of raw sludge. Symbols: ○, acetate; □, propionate; △, butyrate.

increase in concentration as a result of its formation from acetogenic substrates. After 4 hours, the concentration decreased because the rate of acetate dissimilation now exceeded its rate of production. Propionate and butyrate exhibited the same trend but to a lesser extent than acetate. A control vessel with added raw sludge did not exhibit an increase in volatile acids during the same period. Because of the importance of acetate in methanogenesis, factors responsible for its formation were investigated further.

Organisms present in the raw feed sludge might be responsible for the rapid increase in acetate concentration, particularly since their pres-

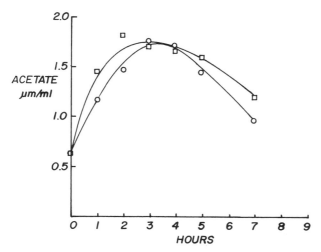

Figure 2.   Acetate production from raw sludge. Symbols: ○, raw sludge; □, sterile raw sludge.

ence is indicated by a concentration of acetate in the raw sludge sufficient to double the acetate level immediately upon feeding. However, Figure 2 shows that the addition of sterilized and unsterilized raw sludge to samples of digesting sludge showed no significant differences in the rates of acetate production from the two sludges. Thus, the conversion of the raw sludge substrates to acetate was accomplished by organisms present in the digesting sludge.

In an effort to establish the nature of the acetogenic substrates, raw sludge was separated into solids and supernatant fractions and were tested individually as substrates in the digesting sludge. Most of the substrates responsible for the rapid formation of acetate were found in the water-soluble solids fraction (Figure 3), and no acetogenic activity was observed in the supernatant fractions.

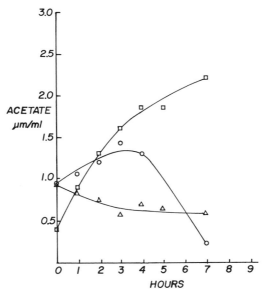

Figure 3. Acetate production from raw sludge fractions. Symbols: ○, raw sludge; □, raw sludge solids; △, raw sludge supernate.

Further fractionation of the solids was effected by extraction with a chloroform–methanol mixture to separate the lipids. The control sludge and the fractions were tested for acetogenic activity at twice their normal equivalent sludge concentrations to ensure production of measurable quantities of acetic acid. The data plotted in Figure 4 indicate that the lipid fraction contains the substrates responsible for the rapid increase in acetate following addition of raw sludge. Acetate increased at approximately the same rate as the raw sludge control. No acetate was observed

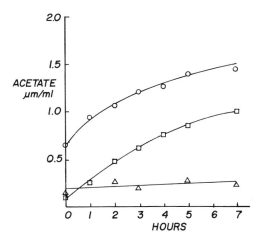

*Figure 4.   Acetate production from lipid
extract of raw sludge. Tested at twice the
normal concentration. Symbols: O, raw
sludge; □, lipid extract; △, residue.*

from the lipid extracted residue. Previous investigations on the composition of sludge indicated that saturated long chain fatty acids, presumably derived from triglycerides, constituted the bulk of the lipid fraction; sterols were present in low amounts and phospholipids were not detected (3, 6, 7, 9). Consequently, chemically pure palmitate was tested as an acetogenic substrate; it was converted to acetate at a rate similar to that of the lipid extract. Acetate was not produced from octanoic acid or tripalmitate, but it was produced more rapidly than other end products from several additional substrates including glycerol, protein hydrolysate, glucose, ethanol, etc. (Table III). The rate of dissimilation of palmitate

**Table III.   Rates[a] of Production of Formate, Acetate, Propionate, and
Butyrate from Several Substrates**

	Production Rate, 10^{-2} μgram C/ml/min				
Substrate	*Formate*	*Acetate*	*Propionate*	*Butyrate*	*Total*
Palmitate	310.0	1420.0	104.0	249.0	2083.0
Protein hydrolysate	19.9	120.0	76.2	20.5	237.0
Pyruvate	34.3	39.7	10.4	4.37	88.8
Glucose	3.97	17.3	9.36	0.83	30.4
Lactate	19.2	0.61	2.16	1.10	23.1
Glycerol	0	14.2	4.45	0.27	18.9
Ethanol	1.96	6.52	4.80	0	13.3
Succinate	1.14	0.64	5.19	0.69	7.66

[a]Calculated from conversion of ¹⁴C-labelled substrate to acid end products.

to acetate was more than 11 times faster than the next best substrate, protein hydrolysate, and more than 80 times faster than glucose or glycerol. Formate, propionate, and butyrate were also produced from these substrates at differing and lower rates. The formate activity shown for pyruvate probably reflects an active phosphoroclastic type reaction, but the activity for lactate is not explainable on the basis of a single reaction. The important conclusion from these data is the significance of palmitate (and probably other long chain fatty acids) as contributors to the acetate pool in the unenriched sludge fermentation.

In all of these experiments, the actively fermenting sludge effects a continuous over-all conversion of the test substrates to CO_2 and CH_4. This implies that acetate is used continuously during the fermentation, and a knowledge of turnover rate is necessary in evaluating acetate formation. Since acetate is presumably metabolized anaerobically only to CO_2 and CH_4, the inhibition of this reaction by the use of methanogenic inhibitors would permit an estimation of the rate of formation of acetate if acetogenesis were unaffected. In this case, turnover of acetate does not occur, and acetate formation can be measured directly. Chlorine derivatives of methane inhibited methanogenesis (24, 25, 26) without affecting the total rate of volatile acid production in the rumen (25) although propionate was higher and acetate lower than normal. We therefore examined CCl_4 and $CHCl_3$ for their effect on methanogenesis in the sludge fermentation. Methane was inhibited 100% by 80 μM CCl_4; however, acetate production was also inhibited at this concentration. Almost complete inhibition of methanogenesis was achieved with as little as 3 μM $CHCl_3$, and at this concentration acetate was apparently produced at its normal rate. Unlike the rumen fermentation, molecular hydrogen was not produced in the presence of these inhibitors.

To assess the utility of inhibitors in measuring acetate production from various substrates, the results obtained by this method were compared with those calculated from turnover studies using radioactively labelled acetate according to the method previously described by Smith and Mah (2). Figure 5 shows the rates of acetate production for sludge determined by these two methods. The slope of the line in the upper figure gives the rate of acetate formation directly as 0.00995 μmole/ml/minute. The slope of the line in the lower graph depicts a turnover rate constant which gives a calculated rate of acetate production of 0.012 μmole/ml/minute. The 19% difference between the two determinations may be partially explained by blockage of the hydrogen-removing methane fermentation which causes the electrons to flow in the direction of other, less adequate terminal hydrogen acceptors. The latter type of fermentation might normally compete with the methane organisms for electrons and would now gain predominance.

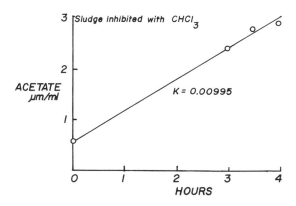

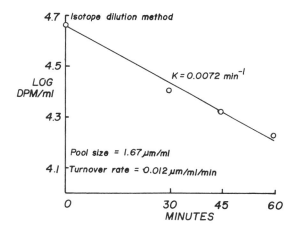

Figure 5.   Acetate production measured by isotope dilution and inhibition by chloroform

Despite the lower rate in the presence of inhibitor, radioactively labelled palmitate was added to a $CHCl_3$-inhibited system. Table IV shows that butyrate was formed at a faster rate than formate, acetate, or propionate. The fact that butyrate was now one of the major end products of palmitate dissimilation indicates that secondary reactions involving acetate and/or propionate were probably serving to remove hydrogen produced during dissimilation since methanogenesis was inhibited in these experiments. This was partially verified by the findings that radioactively labelled acetate was converted to formate and butyrate at faster rates in inhibited than in uninhibited sludge. It is also possible that formation of butyrate indicates some alternative to β-oxidation as a dissimilatory reaction. Acetate itself was formed from $^{14}CO_2$ in the presence or absence of inhibitors and added substrates. The production of acetate from CO_2 has not been reported previously in sludge although it is not surprising that organisms effecting this conversion exist in this environment.

Table IV.   Conversion of Palmitate-U-^{14}C to Volatile Acids in Chloroform-Inhibited Sludge

Acid	Rate of Formation, 10^{-3} $\mu M/ml/min$
Formate	0.079
Acetate	0.080
Propionate	0.414
Butyrate	2.00

Discussion

Various fractions of raw sludge were examined to determine the origin of acetate. We found that the rate of production of acetate from lipid extracts of the solids fraction of raw sludge was approximately equal to that from an equivalent amount of raw sludge. Acetate was not produced from the supernate or non-lipid fraction of the solids. It was concluded that lipids were rapidly metabolized by the sludge population and accounted for the production of gas and volatile acids immediately following addition of raw sludge. Chemically pure palmitate was also metabolized rapidly, and acetate was the main end product. Other substrates, including carbohydrates and proteins, were metabolized more slowly and probably contribute to the gas and volatile acids produced during endogenous metabolism.

The rates of fermentation and degree of dissimilation of various substrates reflect the types of bacteria resident in the particular sludge under study. At any given moment, the population of bacteria depends on a number of factors such as pH, temperature, substrates, oxidation–reduction potential, and retention time. Demonstration of the importance of lipids in this study is an example of the influence of retention time, assuming the other variables were held more or less constant. A long retention time permits development of organisms with long generation times; a short retention time eliminates such organisms. Inhibition of lipid degradation and methanogenesis (9, 27) occurred in lipid-enriched sludge on short retention times; protein- and carbohydrate-enriched sludges were not affected under similar conditions. For this reason, lipids and volatile acids were presumed to be dissimilated at a slower rate than proteins or carbohydrates and limited the minimum retention time required for dissimilation of all degradable substrates. Lipids are, in fact, degraded at higher rates than other substrates examined. Normal sludge retention times in treatment plants are long enough to permit growth of lipid decomposers since lipid dissimilation occurred at rapid rates in all such sludges tested; short retention times result in a wash-out of the slow-growing lipid decomposers.

Mahr (28) and Nikitin (29) reported recently that carbohydrates and proteins are dissimilated more rapidly than lipids by anaerobic sludge. The disparity between these results and ours is apparently caused by the previous enrichment for carbohydrate and protein decomposing organisms in their fermentations. Our experiments are based on short term exposure of unenriched sludge to the substrate being tested. We found that carbohydrates were not rapidly dissimilated in unenriched sludges. Correspondingly, organisms which promote rapid dissimilation of carbohydrates were present only in low numbers (20). However, in the presence of excess substrate, they responded by an energetically wasteful conversion of the carbohydrates to acid end products, and because of their short generation times (20 minutes) they became the most numerous organisms in a very short time (20). Such organisms were now predominantly non-methanogenic euryoxic organisms in contrast to the unenriched population which was composed mainly of strictly anaerobic non-methanogenic *Bacteroides* species (21). Thus, any fermentable carbon and energy source, particularly carbohydrates, can alter the resident population if it is added at high concentrations and is more rapidly metabolized by organisms initially present at low concentrations. In a system where such substrates are continually fed, the heterogeneous mixture of organisms must undergo adjustments in the predominating types and shift to a new equilibrium to maintain the over-all conversion of the starting substrate to methane and carbon dioxide. Failure to do so will result in cessation of methanogenesis.

In contrast to the rumen fermentation, gaseous hydrogen was not produced in the presence of $CHCl_3$ or CCl_4. The chief pathway of methane formation in the rumen is reduction of CO_2 (30) by hydrogen; inhibition of this reaction should lead to the formation of molecular hydrogen. By contrast, most of the methane in the sludge fermentation is formed from acetate (1, 2). Methane can presumably be formed only by direct reduction of the methyl group of acetate (or methanol) (31, 32) or by reduction of CO_2 (33) by hydrogen. Since oxidation of sludge substrates occurred in the presence of methanogenic inhibitors without hydrogen production, some other reactions must have taken place to remove the hydrogen. Our results indicated that several electron acceptors may participate, depending on the nature of the substrate. Radioactively labelled palmitate continued to be oxidized at the expense of some other electron acceptor (probably *via* condensation of two molecules of acetate), and butyric and acetic acids accumulated as end products.

Other terminal electron acceptor products for other substrates were formate, acetate, and propionate. In all of these instances, accumulation of these end products must lead to the eventual inhibition of the fermentation since they are not being converted to CH_4 and CO_2 and consequently removed from the system.

These results are not in agreement with the findings of Thiel (34) who reported the formation of molecular hydrogen in the presence of methane analog inhibitors. However, the organisms examined in his experiments were washed before inhibitors were added, so the conditions were not the same as those reported here. External substrates which may serve as terminal hydrogen acceptors would not be present. The additional use of cells from a laboratory digester maintained on a mixed synthetic substrate may also contribute to the dissimilar findings.

A vigorous methane fermentation requires a balance between the oxidizable substrates and the reduced products. In the sludge fermentation, the oxidizable substrates are organic compounds, and the ultimate reduced product is methane. It is clear that the continued production of methane is a desired feature of the waste treatment fermentation since this gas is easily removed from the system. Reduced end products other than methane are not desirable since such products are mainly soluble organic acids which not only impart noxious odors to the atmosphere but are not easily separated and disposed of. The eventual result is an inhibition of the fermentation arising from increased acid end products which cause a lowering of the pH and/or exert an actual toxic effect on the bacterial cells themselves. Thus, although any number of intermediates can serve as an electron acceptor for oxidation of the organic compounds present in the waste, continued oxidation depends upon removal of the end products. Disturbances in the over-all oxidation–reduction reactions often result in an accumulation of products other than methane, and a cessation of the fermentation ensues.

Thus the heterogeneous population of sludge bacteria is subject to sudden changes in composition as a result of the immediate expression of various selective pressures which may elicit growth of insignificant organisms. Such changes may result in such a drastic shift in the population of organisms that the over-all methane fermentation is inhibited by the accumulation of acid end products. Conversely, inhibition of growth or killing of the sensitive methane organisms by physical or chemical factors may precede the accumulation of acid end products. It is the balanced activity of the mixed population of bacteria which permits operation of the sludge fermentation as a practical process for disposal of organic materials *via* methane formation.

Acknowledgments

This investigation was supported by Public Health Service Grant WP-00921-03.

Literature Cited

(1) Jeris, J. S., McCarty, P. L., "The Biochemistry of Methane Fermentation Using ^{14}C Tracers," *J. Water Pollut. Contr. Fed.* (1965) **37**, 178–192.

(2) Smith, P. H., Mah, R. A., "Kinetics of Acetate Metabolism Drug Sludge Digestion," *Appl. Microbiol.* (1966) **14**, 368–371.

(3) Balmat, J. L., "Chemical Composition and Biochemical Oxidation of Particulate Fractions in Domestic Sewage," Ph.D. Thesis, Rutgers, The State University, 1955.

(4) Buswell, A. M., Neave, S. L., "Laboratory Studies on Sludge Digestion," *Ill. State Water Surv. Bull.* (1930) **30**, 1–84.

(5) Heukelekian, H., "Basic Principles of Sludge Digestion," in "Biological Treatment of Sewage and Industrial Wastes," Vol. 2 "Anaerobic Digestion and Solids-Liquid Separation," J. McCabe, W. W. Eckenfelder, Jr., Eds., pp. 25–43, Reinhold, New York, 1958.

(6) Heukelekian, H., Balmat, J. L., "Chemical Composition of the Particulate Fractions of Domestic Sewage," *Sewage Ind. Wastes* (1959) **31**, 413–423.

(7) Hunter, J. V., Heukelekian, H., "The Composition of Domestic Sewage Fractions," *J. Water Pollut. Contr. Fed.* (1965) **37**, 1142–1163.

(8) Maki, L. R., "Experiments on the Microbiology of Cellulose Decomposition in a Municipal Sewage Plant," *Antonie van Leeuwenhoek J. Microbiol. Serol.* (1954) **20**, 185–200.

(9) O'Rourke, J. T., "Kinetics of Anaerobic Waste Treatment at Reduced Temperatures," Ph.D. Thesis, Stanford University, 1968.

(10) Buswell, A. M., Symons, G. E., Pearson, E. L., "Observations on Two-stage Digestion," *Sewage Works J.* (1930) **2**, 214–218.

(11) Heukelekian, H., Mueller, P., "Transformation of Some Lipids in Anaerobic Sludge Digestion," *Sewage Ind. Wastes* (1958) **30**, 1108–1120.

(12) Rudolfs, W., "Decomposition of Grease During Digestion, Its Effects on Gas Production and Fuel Value of Sludges," *Sewage Works J.* (1944) **16**, 1125–1155.

(13) Sawyer, C. N., Roy, H. K., "A Laboratory Evaluation of High-rate Digestion," *Sewage Ind. Wastes* (1955) **27**, 1356–1363.

(14) Woods, C. E., Malina, J. F., "Glycine Uptake by Anaerobic Waste Water Sludge," *Proc. Purdue Ind. Waste Conf., 19th, Purdue Univ., 1964*, 1011–1024.

(15) Jeris, J. S., Cardenas, P. R., "Glucose Disappearance in Biological Treatment Systems," *Appl. Microbiol.* (1966) **14**, 857–864.

(16) Andrews, J. F., Pearson, E. A., "Kinetics and Characterizations of Volatile Acid Production in Anaerobic Fermentation Processes," *Int. J. Air Water Pollut.* (1965) **9**, 439–461.

(17) Kotze, J. P., Thiel, P. G., Toerien, D. F., Hattingh, W. H. J., Siebert, M. L., "A Biological and Chemical Study of Several Anaerobic Digesters," *Water Res.* (1968) **2**, 195–213.

(18) McCarty, P. L., Jeris, J. S., Murdock, W., "Industrial Volatile Acids in Anaerobic Treatment," *J. Water Pollut. Contr. Fed.* (1963) **35**, 1501–1516.

(19) Toerien, D. F., Siebert, M. L., Hattingh, W. H. J., "The Bacterial Nature of the Acid-forming Phase of Anaerobic Digestion," *Water Res.* (1967) **1**, 497–507.

(20) Mah, R. A., Meck, Lena, Chynoweth, D. P., "Glucose Fermentation by Anaerobic Digester Sludge," unpublished data.

(21) Mah, R. A., Sussman, Carol, "Microbiology of Anaerobic Sludge Fermentation. I. Enumeration of the Non-methanogenic Anaerobic Bacteria," *Appl. Microbiol.* (1967) **16**, 358–361.

(22) Bligh, E. G., Dyer, W. J., "A Rapid Method of Total Lipid Extraction and Purification," *Can. J. Biochem. Physiol.* (1959) **37**, 911–917.

(23) Wiseman, H. G., Irvin, H. M., "Determination of Organic Acids in Silage," *Agr. Food Chem.* (1957) **5**, 213–215.

(24) Bauchop, T., "Inhibition of Rumen Methanogenesis by Methane Analogues," *J. Bacteriol.* (1967) **94**, 171–175.

(25) Rufener, W. H., Jr., Wolin, M. J., "Effect of CCl_4 on CH_4 Volatile Acid Production in Continuous Cultures of Rumen Organisms and in a Sheep Rumen," *Appl. Microbiol.* (1968) **16**, 1955–1956.

(26) van Nevel, C. J., Hendrickx, Demeyer, D. I., Martin, J., "Effect of Chloral Hydrate on Methane and Propionic Acid in the Rumen," *Appl. Microbiol.* (1969) **17**, 695-700.

(27) McCarty, P. L., "Kinetics of Waste Assimilation in Anaerobic Treatment," *Develop. Ind. Microbiol.* (1966) **7**, 144–155.

(28) Mahr, I., "Role of Lower Fatty Acids in Anaerobic Digestion of Sewage Sludge and Their Biocenology," *Water Res.* (1969) **3**, 507–517.

(29) Nikitin, G. A., "Changes in Oxidation-reduction Potential and Intensity of Methane Fermentation During Fermentation of Different Media by a Batch Culture of Methane Forming Bacteria," *Prikl. Biokhim. Mikrobiol.* (1967) **3**, 46 [Engl. transl., *Appl. Biochem. Microbiol.* (1968) **3**, 1].

(30) Hungate, R. E., Smith, W., Bauchop, T., Yu, Ida, Rabinowitz, J. C., "Formate as an Intermediate in the Bovine Rumen Fermentation," *J. Bacteriol.* (1970) **102**, 389–397.

(31) Buswell, A. M., Sollo, E. W., Jr., "The Mechanism of Methane Fermentation," *J. Amer. Chem. Soc.* (1948) **70**, 1778–1780.

(32) Stadtman, T. C., Barker, H. A., "Studies on the Methane Fermentation. VII. Tracer Experiments on the Mechanism of Methane Fermentation," *Arch. Biochem.* (1949) **21**, 256–264.

(33) Barker, H. A., "Bacterial Fermentations," pp. 1–27, Wiley, New York, 1956.

(34) Thiel, P. G., "The Effect of Methane Analogues on Methanogenesis in Anaerobic Digestion," *Water Res.* (1969) **3**, 215–223.

RECEIVED August 3, 1970.

Toxicity, Synergism, and Antagonism in Anaerobic Waste Treatment Processes

IRWIN JAY KUGELMAN[1]

American-Standard, Inc., New Brunswick, N. J.

K. K. CHIN

Singapore Polytechnic, Singapore 2

Antagonism, synergism, and complexing reactions must be considered in reporting data on toxicity and stimulation in anaerobic waste treatment systems. Low concentrations (< 1 mg/liter) of all heavy metals except iron are extremely toxic, but higher concentrations can be tolerated if sufficient sulfide is present to act as a precipitant. Short chain volatile acids up to 6000 mg/liter have no adverse effect on methane formers. While pH control with alkaline substances can prevent catastrophic failure during digestion, the fundamental cause of imbalance must be corrected for a permanent cure. Light metal cations offer the most difficulty in designing and operating these systems because they can occur in both toxic and stimulatory ranges. An initial attempt to categorize quantitatively toxicity, stimulation, etc. on an absolute basis is presented.

A dequate knowledge of toxicity phenomena is crucial to the proper design and operation of a biological waste treatment system. This is especially true for the anaerobic waste treatment process because the key group of bacteria in the process, "the methane bacteria," is much more sensitive to environmental conditions than other groups of bacteria. Although this process has significant advantages over other methods of waste treatment, its use has been retarded because of a lack of under-

[1] Present address: Advanced Waste Treatment Program, Environmental Protection Administration, R. A. Taft Sanitary Engineering Center, 4676 Columbia Parkway, Cincinnati, Ohio 45226.

standing of toxicity phenomena. There are many reports in the sanitary engineering literature which detail the failure or inefficient operation of this process. In many situations, it is apparent that a toxicity situation existed but was not considered in the failure mode analysis. More frequently, toxicity was suspected, but the wrong substance was indicated as the toxic factor (1, 2, 3).

During the last 10 years, much fundamental information has been discovered concerning anaerobic waste treatment. With judicious use of this information, anaerobic waste treatment systems can be designed and operated in a less empirical and more efficient manner than previously. Included in these fundamental findings are basic data on toxicity. This paper presents a review of the available information on toxicity in anaerobic waste treatment. The data have been analyzed and are presented from the viewpoint of the waste treatment engineer. This mode of presentation was chosen to facilitate the utilization of the data presented and thus expand the applicability of the anaerobic waste treatment process.

Fundamental Considerations

The basic information which the sanitary engineer requires is a quantitative evaluation of toxicity under any set of environmental conditions. As discussed below, it is difficult to arrive at such an evaluation. The magnitude of the toxicity observed or reported is a function of several factors including concentration, antagonism, synergism, complex formation, and acclimation. Each of these is discussed in turn.

Concentration. It is usual to classify qualitatively a substance as either toxic or stimulatory. For example, heavy metals are considered to be toxic, while glucose is classified as a food or growth stimulant. As Figure 1 illustrates, such a view is somewhat simplistic in that the effect of any substance on the metabolism of an organism is concentration dependent. As the concentration of any metabolite increases, the metabolic rate first increases, enters a constant zone, and eventually declines. At very low and very high concentrations the metabolic rate is negligible. Thus speaking absolutely, a substance is never toxic, but its degree of metabolic stimulation is concentration dependent.

In general, when a substance is in the concentration range below its peak range, it is considered to be in its stimulatory range; when the concentration is above its peak range, the substance is considered to be in its toxic range. These definitions must be termed "common usage" because the metabolic rates obtained in the stimulatory region are not higher than those obtained in the toxic region. What is actually being defined is the trend of the metabolic rate with concentration increase.

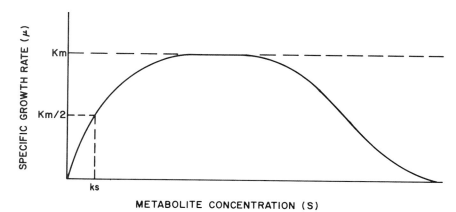

Figure 1.   *Effect of metabolite concentration on growth rate*

Whether or not a substance is classified by common usage as a food or a toxin depends on the position of its usual concentration range with respect to the range of peak stimulation. Heavy metals are considered toxic because they can be present in wastes at a concentration of a few milligrams/liter, which is several orders of magnitude above their range of peak stimulation (4). Glucose is considered a stimulant because it is usually present in wastes at a concentration of several hundred milligrams/liter, which is well below its range of peak stimulation.

Just as qualitative judgments of toxicity are made using a reference point on the metabolism rate–concentration relationship, quantitative evaluation of toxicity is also made with respect to a reference point. It would seem that the definition of the point of reference for quantitative evaluation should be more exact than that used for qualitative judgment. Unfortunately, this is not the case for waste treatment system studies. In virtually all quantitative toxicity studies, the performance of units containing added quantities of the substance under study is compared with the performance of a control unit. Invariably, the control unit is one which exhibits satisfactory biological activity and to which none of the substance under study has been added. Very rarely is the concentration of the substance under study established or fixed in the control.

Without establishing a standard control, results from study to study are virtually impossible to compare. Indeed, this is one reason why quantitative toxicity studies in waste exhibit such conflicting results.

It would be best if toxicity data were put on an absolute rather than a relative basis, since reference to a control unit would not be required. The advantage of defining quantitatively the metabolic rate–concentration curve (Figure 1) over the whole or most of its range is obvious. The initial attempts to do this for light metal cations are discussed later.

Antagonism and Synergism. In dealing with toxicity, antagonism and synergism are most important. Antagonism is a reduction of the toxic effect of one substance by the presence of another. Synergism is an increase in the apparent toxicity of one substance caused by the presence of a second substance in the environment. As with toxicity, antagonism and synergism are concentration dependent. Figure 2 illustrates typical antagonistic and synergistic patterns in two-component systems. In this system, the concentration of A is held constant at a concentration level which is defined as toxic. With the concentration of B at zero, the metabolic reaction rate is represented by the point on the ordinate axis. If B is an antagonist, the reaction rate will increase to a peak and then decline as the concentration of B is increased. At some concentration of B, called the crossover concentration, the reaction rate will equal that obtained when the concentration of B was zero. The crossover concentration marks the point at which the antagonistic action of B terminates. If B acts as a synergist, the reaction rate never increases but starts to decrease with any increase in the concentration of B.

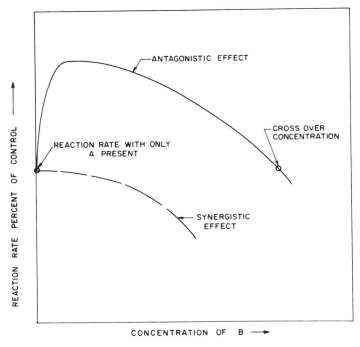

Journal of the Water Pollution Control Federation

Figure 2.   Antagonistic and synergistic relationships in two-component systems (23)

Antagonism and synergism may be manifest between substances which have a chemical relationship, such as light metal cations (5), or between substances as disparate as antibiotics and the hydrogen ion (6). Thus, qualitative prediction of the presence of antagonism or synergism is quite difficult. On a quantitative level, the most significant aspect of these phenomena is that only small concentrations of the antagonist and/or synergist are required to produce a significant change in the reaction rate.

Therefore, to define the toxicity of any substance, it is necessary to delineate the substances which can exert antagonistic or synergistic effects and the concentration ranges in which these are manifest. Unfortunately, complete information on antagonism and synergism for most substances is not available.

Complex Formation. If a substance is not in solution, it cannot gain entrance to the cell and thus cannot affect the metabolism of an organism. Thus, complex formation (*i.e.*, removal of a substance from effective solution) has a significant effect on the toxicity observed in any situation. It is most important in any toxicity situation to understand the potential of the substances present to partake in complex formation reactions. In some situations, complex formation can be predicted easily on the basis of the substances present. In other situations, complex formation is difficult to gage since it will result only after biological action alters some of the components.

Acclimation. The magnitude of the toxic effect generated by a substance can often be reduced significantly if the concentration is increased slowly. This is the phenomenon of acclimation which represents an adjustment of the biological population to the adverse effects of the toxin. In a waste treatment system this does not usually represent either a mutation or a selection of the fittest because rarely is all of the toxic effect eliminated. Rather acclimation represents a rearrangement of the metabolic resources of the organism to overcome the metabolic block produced by the toxic substance. Significant reductions in toxicity may be obtained if the concentration of the toxic substance is increased slowly rather than increased all at once.

In most toxicity studies, the concentration of the toxic substance is raised in one step to the level to be studied, leaving little chance for acclimation to take place. This experimental situation is not adequate for toxicity studies in waste treatment systems because in most of these systems acclimation will probably occur. This is true because if a toxic substance is present in a waste, its concentration in the reactor will usually build up slowly rather than suddenly. Thus, adequate delineation of toxicity in waste treatment requires an allowance for acclimation.

Toxicity in Anaerobic Waste Treatment

A review of the literature on anaerobic waste treatment indicates considerable variation in the toxicity reported for most substances. In some cases, the range is over more than one order of magnitude. The major reason for these variations is the complexity of the phenomenon of toxicity. Unfortunately, in most studies of toxicity in anaerobic waste treatment little or no attention is paid to the effect of antagonism, synergism, acclimation, and complexing reactions. Indeed, in very few studies are the chemical characteristics of the control defined adequately.

The main reasons for these experimental deficiencies are:

(1) Much published toxicity data for anaerobic waste treatment are the product of studies in which toxicity was not to be studied! Rather, toxicity developed during a study of some other phenomenon, and the toxicity data were reported, sometimes as an afterthought, with the other data.

(2) Another main source of toxicity data is from reports on failure of and attempts to revive field digestion systems. These reports were primarily filed by treatment plant operators who were not equipped to run the sophisticated chemical analyses required to categorize adequately the environment in the digester. Neither for that matter did they have the time since coping with a field scale "sour" digester is a formidable task.

(3) Most studies have been conducted in full strength or diluted sewage sludge which is an incredibly complex substance whose characteristics vary with geographical location. This procedure has been necessary because it is difficult to develop a chemically defined medium in which methane bacteria will flourish. These organisms will thrive in sewage sludge.

Although there are good reasons why much of the data on toxicity in anaerobic systems are suspect, such data have not and cannot be interpreted properly. At best, most of these data are useless; at worst, they are misleading. Recently, studies of toxicity in anaerobic waste treatment were conducted utilizing experimental procedures which circumvent the inadequacies discussed above. Emphasis is placed on the data from these studies in the following sections of this paper.

Figure 1 shows that any substance can exert a toxic effect, and many have been reported as toxic in anaerobic treatment. However significant data exist for only three groups of substances—heavy metals, organic acids (pH), and light metal cations.

Heavy Metals

Heavy metals have frequently been the cause of poor performance of biological waste treatment units because of their extreme toxicity. For example, Hotchkiss (4) reported that E. coli would not grow when the

Table I.   Reported Values of Toxic Concentrations of Heavy Metals in Anaerobic Waste Treatment

Metal	Toxic Concentration, mg/liter	Reference
Copper	150–250	30
	500	31
	1000	32
Nickel	200	32
	1000	33
Zinc	1000	34
	350	35
Chromium	2000	32
	200	36

concentration of zinc, lead, nickel, chromium, iron (ferric), tin, cadmium, or cobalt exceeded the range 10^{-3}–$10^{-4}M$. Adequate growth took place only when the concentration range was reduced to 10^{-5}–$10^{-6}M$. Reported concentrations at which toxicity occurs in anaerobic waste treatment systems, however, are much higher than the range above. In addition, the variation in reported values is considerable. Table I summarizes representative data. The high tolerance levels and variability reported are probably caused by the ease with which heavy metals take part in complex-type reactions with the normal constituents of an anaerobic waste treatment unit. Examples of these are precipitation by sulfides and sequestering by ammonia and reactive groups of organic material present.

Evidence of the effectiveness of complex-type reactions in reducing the concentration of heavy metals in anaerobic waste treatment is illustrated in Table II. These data were taken from a study by Barth *et al.* (7) on the effect of heavy metals on all types of biological waste treatment systems. The metal concentration in solution was reduced by a factor of over 100 by complexing reactions. Table III, also taken from this study, gives the concentration at which each of several heavy metals can be present in sewage without adversely affecting anaerobic digestion of the resulting sewage sludge. These results take into account the complex-type reactions which can be expected under anaerobic conditions.

Table II.   Total and Soluble Heavy Metal Content of Digesters (7)

Metal	Total Concentration, mg/liter	Soluble Concentration, mg/liter
Chromium (VI)	420	3
Copper	196	0.7
Nickel	70	1.6
Zinc	341	0.1

**Table III.   Highest Dose of Metal that Will Allow Satisfactory
Anaerobic Digestion of Sludge's Continuous Dosage (7)**

Metal	Concentration in Influent Sewage, mg/liter	
	Primary Sludge Digestion	Combined Sludge Digestion
Chromium (VI)	>50	>50[a]
Copper	10	5
Nickel	>40	>10[a]
Zinc	10	10

[a] Higher dose not studied.

The most important complex-type reaction for controlling toxicity in anaerobic waste treatment is the precipitation of heavy metals by sulfides. This was noted by Barth *et al.* (7) and Masselli and Masselli (8). The reason for the preeminence of sulfides is the extreme insolubility of heavy metal sulfides. The solubility product of heavy metal sulfides ranges from 3.7×10^{-19} for FeS to 8.5×10^{-45} for CuS (9). Thus, in the presence of sulfides, the heavy metal concentration is expected to be virtually zero in an anaerobic waste treatment unit.

Lawrence and McCarty (10) conducted an extensive investigation of the ability of sulfides to prevent heavy metal toxicity. They added 13.5 mM and 6.25 mM concentrations of several heavy metal sulfates to laboratory scale sewage sludge digestors. Although the heavy metal concentration reached several hundred milligrams/liter, performance of all units was satisfactory because the sulfate was broken down biologically to sulfides, which in turn precipitated the heavy metals. After 57 days of operating a chloride heavy metal salt was substituted for the sulfate salt in the 6.25 mM units, but sulfate feed was maintained in the 13.25 mM units. Almost immediately gas production started to decrease in the units where sulfate was discontinued except for the unit which was fed iron. In all units where sulfate feed was maintained, good performance was maintained. Figure 3 illustrates these results, showing that within a few days after sulfate feed was stopped to the copper, nickel, and zinc feed units the gas production had fallen practically to zero. The unit which was fed iron did not fail because at the normal pH of a digester the iron is precipitated as a hydroxide and because iron is much less toxic than other heavy metals. Figure 4 presents the results of studies in which combinations of heavy metals were used (10). Here results virtually identical to those discussed above were obtained. With sulfate in the feed, no toxicity was evident; when sulfate feed was discontinued, severe toxicity was soon evident. Again, iron was found to be non-toxic.

The above results indicate clearly that heavy metal toxicity can be

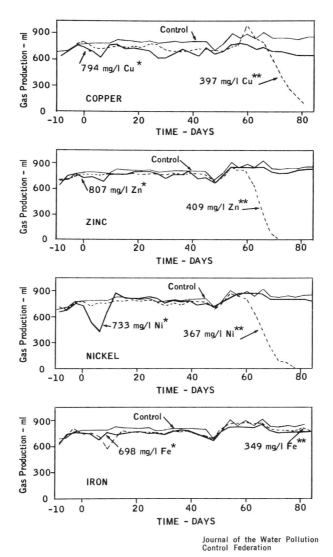

Journal of the Water Pollution Control Federation

Figure 3. Effect of heavy metals on digestion (10)
** Changed from sulfate salt to chloride salt on day 57
* Added as sulfate salt during period shown

eliminated by precipitation with sulfides. The approximate dose required to react stoichiometrically is 0.5 mg/liter of sulfide per 1 mg/liter of heavy metal. These investigators successfully controlled heavy metal toxicity in a field scale digester at the Palo Alto treatment plant with additions of sodium sulfide. However, because high concentrations of sulfide are toxic

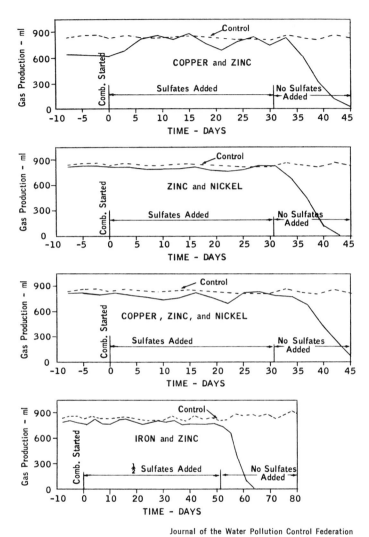

Journal of the Water Pollution Control Federation

Figure 4.   Effect of mixtures of heavy metals on digestion (10)

and can produce severe corrosion problems in digester gas piping, this procedure is not recommended. It was suggested that ferrous sulfate be added continuously to a digester as the source of sulfides. Sulfides will be produced from the biological breakdown of sulfate, and the excess will be held out of solution by the iron. However, if heavy metals gain entrance to the digester, they will draw the sulfide preferentially from the iron because iron sulfide is the most soluble heavy metal sulfide.

Lawrence and McCarty (*10*) also found that when toxicity was manifest, gas production fell off much more rapidly than volatile acids increased. This is indicative of equal toxicity to the acid formers and methane formers. This is unusual in that the methane formers are usually considered to be more sensitive than acid formers to environmental conditions. It is probable that the very high toxicity of heavy metals results in virtual elimination of both groups of organisms once a few milligrams/liter excess of heavy metal are present. This is probably also the reason it is impossible to discriminate between the toxicity of individual heavy metals.

pH and Volatile Acids

Practical experience and research indicated early that anaerobic waste treatment proceeded at its optimum only in a narrow pH range. It is now realized that this is because the methane bacteria flourish only at a pH range of 6.4–7.5 (*11*). Although the need for pH control was recognized, the best or correct procedure for achieving this has been the source of much debate.

Under balanced digestion conditions, the biochemical reactions tend to maintain automatically the pH in the proper range. Although volatile organic acids produced during decomposition of complex organics tend to reduce the pH, this effect is counteracted by destruction of volatile acids and reformation of bicarbonate buffer during methane fermentation. However, if an imbalance develops, the acid formers outpace the methane formers, and volatile organic acids build up in the system. If balanced digestion is not restored, the buffer capacity is overcome, and a precipitous drop in pH occurs. The low pH will stop methane production almost completely but will hinder acid formation only slightly. Thus a "stuck" or "sour" digester results. Restoration of balanced conditions in a "sour" digester requires considerable time because of the low growth rate of methane bacteria. In many cases, balance is achieved only by seeding heavily with digested sludge.

Therefore, a prime objective in the operation of anaerobic waste treatment processes is maintenance of a proper pH range. The signal that trouble is imminent is a sudden rise in the volatile acids. One group of investigators has indicated that the proper action to take when the volatile acids rise suddenly is to add alkaline substances to maintain the buffer capacity (*12, 13, 14*). A second group led by Buswell (*15*) and Schlenz (*16*) contends that this is detrimental because the volatile acids themselves are toxic to methane bacteria but not to acid formers at concentrations of about 2000 mg/liter. The use of alkaline materials thus only stimulates acid production leading to even greater toxicity. This latter group feels that the only proper course of action is to reduce the feed rate and dilute the digester contents. In this manner, the concentration of the "toxic" volatile acids will be reduced, and balanced digestion can be restored.

Until a decade ago, the available experimental data could be used to support either point of view. The major reason was that carefully controlled studies had not been conducted to prove or disprove either contention. At that time, McCarty and McKinney (17, 18) published the results of their work in which they investigated the toxicity of acetic acid to methane bacteria. Acetic acid is the most prevalent acid produced during anaerobic waste treatment (19).

Figure 5 (17) illustrates that acetate is the least toxic anion of any of those commonly found in wastes. However, at the levels used (2000–4000 mg/liter) which are levels reported in failing or stuck digesters,

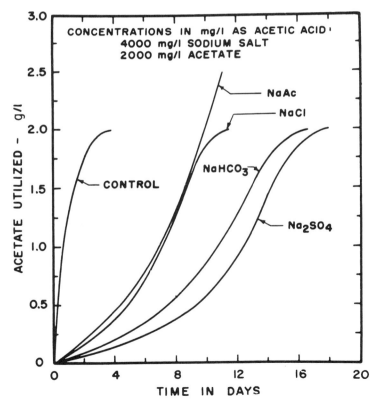

Journal of the Water Pollution Control Federation

Figure 5. Effect of sodium salts on methane production from acetate (17)

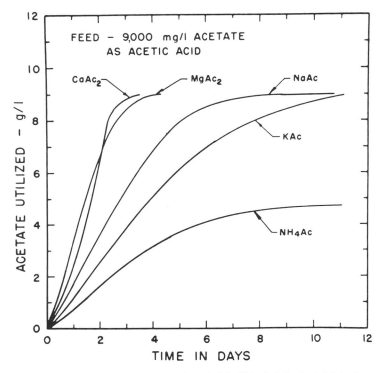

Figure 6.   Effect of acetate salts on methane production (18)

toxicity was evident. These investigators found, however, that the toxicity was caused by the sodium cation not the acetate anion. Figure 6 (*18*), illustrates virtually no toxicity at acetate levels of 9000 mg/liter when calcium or magnesium acetate was fed but toxicity when sodium, potassium, or ammonium acetate was fed. Thus, McCarty and McKinney indicated that pH control by adding alkaline material was a valid procedure provided the cation of the alkaline material did not itself produce toxicity. They suggested the use of magnesium and/or calcium alkaline compounds rather than monovalent alkaline compounds.

This work, however, did not completely eliminate the controversy over the toxicity of volatile acids. Subsequently, Buswell and Morgan (*20*) reported that propionic rather than acetic acid was toxic to the methane bacteria. McCarty *et al.* (*9*) investigated the effect of various volatile acids on methane bacteria in order to clear up this controversy. They added 6000 mg/liter of acetic, propionic, and butyric acid (the three most common acids produced during anaerobic breakdown of complex substances) individually to laboratory scale sewage sludge digesters. Prior to the slug of each acid, the units had been fed sewage sludge daily. Subsequent to the acid slug, daily additions of raw sewage sludge were maintained. The acids were neutralized with non-toxic concentrations of alkaline substances before being fed so as not to affect adversely the pH in the digesters. The results are illustrated in Figure 7. The acetic and butyric acids were metabolized rapidly since their concentration was reduced much more rapidly than if washout with no metabolism occurred. However, metabolism of propionic acid was delayed for about 10 days. A more extensive study of propionic acid was undertaken, utilizing three levels of this acid with the same experimental technique as before. As Figure 8 illustrates, at the highest level used, 8000 mg/liter, some retarda-

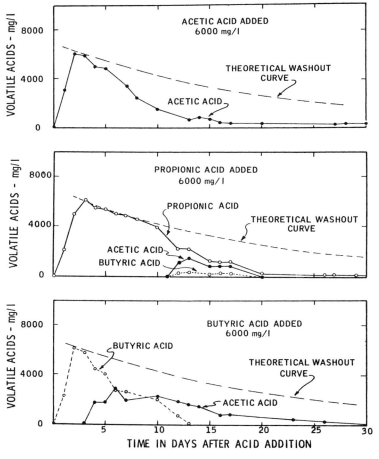

Figure 7.   Effect of slug feed of volatile acids on daily feed sludge digesters

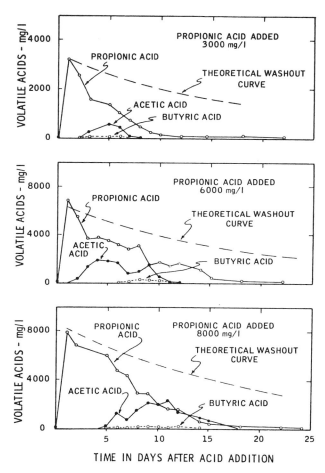

Figure 8.   Effect of slug feed of propionic acid on daily feed sludge digesters

tion of propionate metabolism took place. However, acetic acid did not increase in the medium until after propionate metabolism started.

These data indicated several possibilities concerning propionate inhibition which can be explored with the aid of Figure 9. If propionic acid affected the breakdown of acetic acid to methane and carbon dioxide, acetic acid would have increased in concentration, but this did not occur. If propionic acid affected the breakdown of propionate to acetic acid, propionic acid content would have increased, but it did not. A third possibility is that the propionate slowed the breakdown of complex organics to propionic acid and acetic acid but did not affect the methane formers. Since the data tentatively supported this possibility, an additional run was conducted. One unit was slugged with 6000 mg/liter propionate, a second with 6000 mg/liter propionate plus 2000 mg/liter acetate, and a third with 2000 mg/liter acetate. Figure 10 illustrates that although propionate metabolism was delayed, the acetate was used up rapidly, indicating that propionate had little effect on methane production but did affect acid production adversely.

The studies of McCarty and co-workers have shown clearly that volatile acids are not toxic to methane bacteria at concentrations that would occur in stuck or sour digesters. On the contrary, evidence has been elucidated which indicates propionate retards the acid formers. Thus, the use of alkaline substances to maintain an adequate buffer capacity in an anaerobic waste treatment unit is a valid procedure. A word of caution is necessary; pH control is not a universal palliative. Its only advantage is to prevent a bad situation from getting out of hand. The basic cause of the digester biochemical imbalance must be discovered and rectified. Unless this is done, pH control is worthless in the long run. In addition, care must be exercised in selecting an alkaline material that will not produce a toxic reaction.

Light Metal Cations

The light metal cations are the most interesting and important metabolites covered in this report. As indicated, choice of the alkaline material to control pH hinges on the toxic nature of the cation. In addition, many industrial wastes contain high concentrations of light metal cations which could adversely affect treatability. Of most significance, however, is that the concentration range in which light metal cations occur in most wastes extends from the stimulatory to the toxic section of the rate of metabolism concentration curve of Figure 1. Thus, both stimulation and toxicity may be experienced with light metal cations.

Little work of significance was performed on the effect of light metal cations until the last decade. The early studies such as those of Buswell et al. (3), Rudolphs and Zeller (21), Fair and Carlson (13), Cassell and

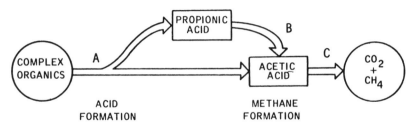

Figure 9.   General pathways of anaerobic degradation of complex wastes

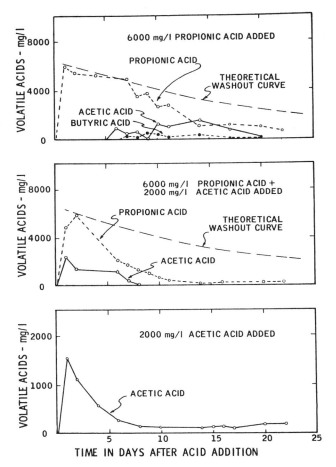

*Figure 10.   Effect of slug feed of propionic and acetic
acids on sludge digestion*

Sawyer (*12*) were conducted in sewage sludge digesters. The ionic constitution of the medium was not defined completely nor controlled. Even the work of McCarty and McKinney (*17, 18*) which first indicated the importance of light metal cation toxicity, was conduced in diluted sewage sludge. In this light, it is significant that many of these early studies reported the divalent cations magnesium and calcium were less toxic or equal in toxicity on a molar basis to the monovalent cations sodium, potassium. This is at variance with all previous experience in toxicity studies in pure biology. For example, Winslow and Haywood (*22*) found calcium and magnesium 10 times as toxic on a molar basis as potassium to *E. coli*.

Table IV.   Growth Medium Utilized for Study of Cation Effects on Acetate-Utilizing Methane Bacteria

Chemical	Concentration, mg/liter
Cambridge tap water	1000
Bicarbonate buffer	2500 as $CaCO_3$
$FeCl_3$	60
$CoCl_2 \cdot 6H_2O$	8
Thiamine–HCl	2
K_2SO_4	10
NH_4Cl	147
$(NH_4)_2HPO_4$	22

At this juncture, Kugelman and McCarty (23) investigated the effects of light metal cations on methane bacteria in which a chemically defined medium was used. The medium they developed is given in Tables IV and V. Although it is low in all inorganic and organic growth factors, it could support acetate-utilizing methane bacteria at a feed rate 0.5 gram/liter/day of acetate. Using this medium, they studied the toxic effect of the five common light metal cations on acetate-utilizing methane bacteria. These organisms were chosen because over 70% of the methane produced in anaerobic treatment of complex wastes is from fermentation of acetate (19), and because all earlier work indicated that methane producers were much more sensitive to toxic effects of light metal cations than acid formers.

Figure 11 illustrates the results of slug feed toxicity studies in systems where the concentration of only one cation was increased above the base level in the growth medium. The results agree more with those from the pure bacterialogical field than those from the sanitary engineering field. The investigators felt that much of the deviation in previous studies was possibly caused by antagonism and synergism. Consequently, dual cation systems were studied to elucidate these relationships.

It was found that either antagonism or synergism was manifest in each system. The phenomena were mutually exclusive: if one occurred with a cation pair, the other did not. Table VI is a summary of data for all systems exhibiting antagonism. The most significant points are:

(1) In several instances not only was antagonism observed, but stimulation as well. For example, while 0.075M magnesium alone re-

Table V.   Ionic Characteristics of Cambridge Tap Water

Ion	Concentration, mg/liter
Sodium	10
Potassium	1.8
Ammonium	<1
Magnesium plus calcium	50–60 as $CaCO_3$

duced the reaction rate to 55% of the control, 0.075M magnesium plus 0.01M sodium produced a reaction rate equal to 113% of the control.

(2) Only very low concentrations of the antagonist were needed to produce significant antagonism. In many cases the peak antagonism range started at an antagonist concentration of 0.002M.

(3) The concentration range of the antagonist at which peak antagonism was achieved narrowed as the concentration of the toxic cation increased. When the peak antagonism range narrowed, the shift was confined almost wholly to the high side of the peak antagonism range. The trend of the narrowing indicates that the focus of the peak antagonism range is 0.01M with monovalent cations as antagonists, and 0.005M when divalent cations serve as the antagonist.

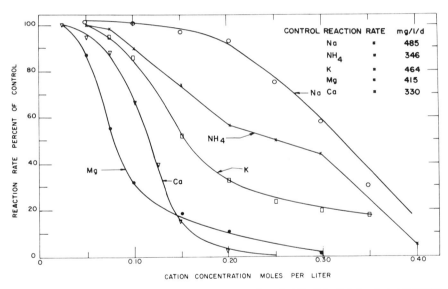

Journal of the Water Pollution Control Federation

Figure 11. Effect of individual cation concentration on rate of acetate utilization (10)

(4) The crossover concentration remained constant or decreased as the concentration of the toxic cation increased.

Table VII summarizes the results for the synergistic systems. The most significant points are:

(1) The concentration of the synergist at which the synergistic effect began was below the concentrations at which the synergistic cation would produce inhibition when present alone. For example, calcium produced a synergistic effect with ammonium at calcium concentrations starting at 0.01M. In a single cation system, however, calcium did not produce any inhibition until its concentration reached 0.05M.

(2) Mutual antagonism between both cations in a cation pair was not observed in many cases. Mutual antagonism is a situation where each cation of a cation pair can antagonize the toxicity produced by the other. For example, in the ammonium–sodium system antagonism was observed with ammonium as the toxic cation while synergism was observed when sodium served as the toxic cation.

The most significant finding of this phase of the study was that very low concentrations of the cations studied could produce both synergistic and antagonistic effects. Thus it is clear why previously reported results on cation toxicity in this field have differed qualitatively from those obtained by biologists. Undoubtedly, since strict control over the ionic constituents of media was not exercised, synergistic and antagonistic effects were overlooked.

Table VI.   Antagonistic Relationships

Toxic Cation	Concentration, M	Reaction Rate, % of Control with No Antagonist Present	Antagonistic Cation
Na	0.3	54	K
Na	0.4	30	K
NH_4	0.15	80	Na
NH_4	0.25	55	Na
K	0.15	40	Na
K	0.25	20	Na
K	0.35	15	Na
K	0.15	41	Ca
K	0.25	20	Ca
K	0.35	16	Ca
K	0.15	43	Mg
K	0.25	20	Mg
K	0.15	41	NH_4
K	0.15	24	NH_4
Mg	0.075	54	Na
Mg	0.1	34	Na
Mg	0.125	13	Na
Mg	0.2	5	Na
Mg	0.075	58	K
Mg	0.1	38	K
Mg	0.125	20	K
Mg	0.2	8	K
Ca	0.1	58	Na
Ca	0.1	55	K

Kugelman and McCarty (23) also investigated multiple cation systems. Because of the immense number of possible combinations, this phase of the work was limited to systems where one cation was raised to a toxic level while the concentration of the other cation was held in the range 0.005–0.01M. This range coincides with the usual range for peak antagonism. The major points determined in these studies were:

(1) Antagonism of toxicity by multiple cations was superior to that achieved by a single antagonist. Figure 12 shows that the presence of sodium plus potassium eliminated more of the toxicity of magnesium than either one alone.

(2) Multiple antagonism does not require that all of the antagonsists be able to antagonize individually the toxic cation. Figure 13 shows that

in Dual Cation Systems

Range of Peak Antagonism, M	Reaction Rate, % of Control with Peak Antagonism	Crossover Concentration, M
0.002–0.06	72	0.1–0.11
0.005–0.03	56	0.11–0.12
0.002–0.05	105	0.08
0.005–0.025	67	0.045
0.005–0.1	84	0.1
0.01–0.05	75	0.1
No antagonism	—	—
0.002–0.05	78	0.1
0.005–0.05	59	0.1
0.002–0.01	38	0.05
0.002–0.075	85	0.075
0.005–0.05	64	0.07
0.05–0.1	60	0.1
0.05–0.1	49	0.1
0.005–0.035	113	0.105
0.005–0.035	103	0.1
0.005–0.025	80	0.1
No antagonism	—	—
0.002–0.06	93	0.12
0.005–0.06	73	0.12
0.005–0.045	62	0.12
No antagonism	—	—
0.005–0.025	94	0.045
0.002–0.06	73	0.12

Table VII. Synergistic Relationships in Dual Cation Systems

Toxic Cation	Concentration, M	Synergistic Cation	Concentration of Synergist where Synergism Begins, M
Na	0.3	NH$_4$	0.025
Na	0.4	NH$_4$	0.01
Na	0.3	Ca	0.01
Na	0.4	Ca	0.05
Na	0.3	Mg	0.01
Na	0.4	Mg	0.05
NH$_4$	0.15	K	0.025
NH$_4$	0.25	K	0.025
NH$_4$	0.15	Ca	0.02
NH$_4$	0.25	Ca	0.01
NH$_4$	0.15	Mg	0.01
NH$_4$	0.25	Mg	0.005
Mg	0.075	NH$_4$	0.01
Mg	0.125	NH$_4$	0.005
Mg	0.075	Ca	0.005
Mg	0.125	Ca	0.005
Ca	0.1	NH$_4$	0.05
Ca	0.1	Mg	0.002

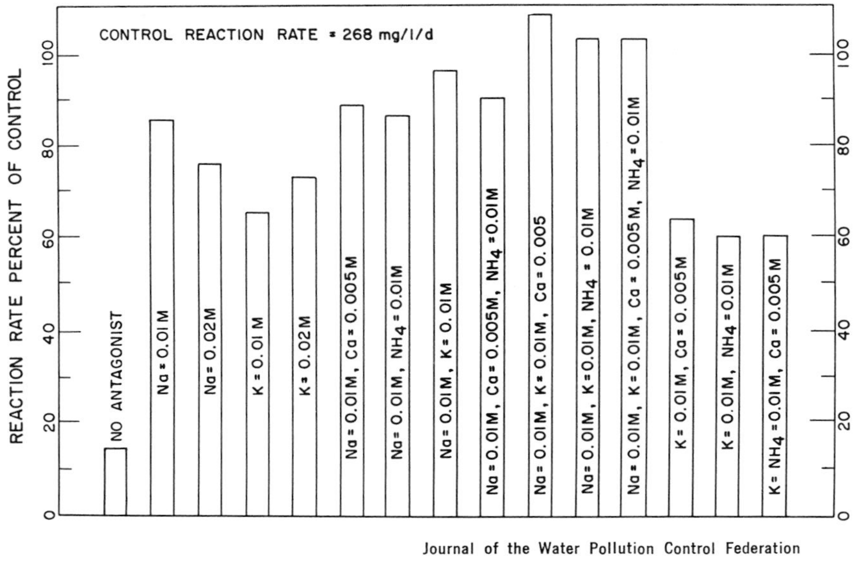

Figure 12.   Antagonism of magnesium by combinations of cations (10)

potassium plus calcium yielded superior antagonism to that achieved by potassium alone even though Tables VI and VII indicate that potassium is an antagonist of sodium but calcium is a synergist. This phenomenon is termed secondary antagonism. Secondary antagonism is also illustrated in Figure 12 in that the presence of calcium and ammonium increases the antagonism of magnesium toxicity achieved by the combination of sodium plus potassium.

(3) Stimulation of metabolism to a rate higher than the control is achieved by multiple cation antagonism even in the presence of a very high concentration of a toxic cation.

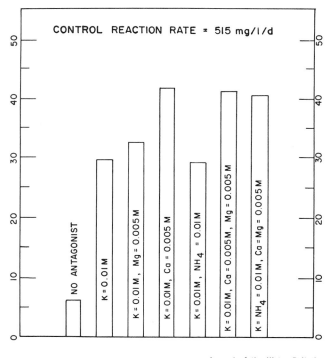

Journal of the Water Pollution Control Federation

Figure 13. Antagonism of sodium by combinations of cations (10)

In addition to yielding fundamental toxicity data and delineating the cation pairs which exhibited antagonism and synergism, the most important finding of this investigation was the relationship between nutrition, stimulation, and antagonism. Antagonism of toxicity was so complete in some instances that metabolic rates higher than the control were obtained. In addition, the concentration of the antagonist which produced peak antagonism remained the same independent of changes in the concentration of the toxic cation. These findings indicate that antagonism is more the result of a stimulatory effect of the antagonist than a direct counter effect to the toxin. Consequently, the concentration of an antagonist which produces peak antagonism must bear a relationship to the maximum nutritional requirement for the antagonist. In short, the concentrations of cations which produce peak antagonism 0.01M for monovalent cations and 0.005M for divalent cations also ensure optimum metabolic activity even when toxic conditions do not exist. To achieve optimum performance of an anaerobic waste treatment unit, the concentration of the light metal cations should be maintained as close as possible to the values above.

The results above were determined in slug feed systems. These data are thus applicable to situations where the cation or cations are suddenly introduced into the waste treatment system. For the most part this is not what one would expect for a toxicity situation in waste treatment. Usually the toxic material will be present in the waste in a continuous or semi-continuous basis. Its concentration will slowly rise to and remain at some equilibrium value in the waste treatment unit. Such conditions allow for acclimation to be manifest. To evaluate cation effects for such situations, Kugelman and McCarty (24) extended their investigations to daily feed units. The same growth medium was used as for the slug feed runs. The daily feed units were operated at a 15-day hydraulic and biological solids retention time at an organic loading of 0.5 gram per liter per day of acetate. Once-daily feed and withdrawal of mixed liquor were used. The major findings of this study were:

(1) The antagonistic and synergistic relationships determined in the slug feed studies were confirmed in the daily feed studies.

(2) Acclimation to the toxic effect of high concentrations of the light metal cations was significant. Figure 14 illustrates the extent of the acclimation observed.

(3) The combination of acclimation and antagonism yielded results superior to either phenomenon alone as a method of curtailing toxicity.

Based on the results of the two studies summarized above, Kugelman and McCarty suggested design values for the maximum cation concentration which could be tolerated in anaerobic waste treatment. These values which are given in Table VIII were chosen so a reaction rate of at least 90% of the control would always be obtained.

The results reported by Kugelman and McCarty represent a significant advance in knowledge of the effects of light metal cations on anaerobic waste treatment, especially in the areas of antagonism, synergism, and delineation of optimum ionic environment. However, the results reported in Table VIII are of limited value to design engineers because they are applicable for only one organic loading, one biological solids retention time, and one definition of non-toxic operation. (If the reaction rate is

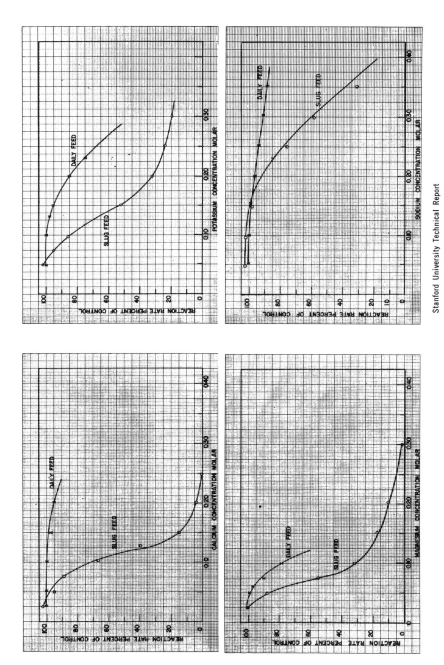

Stanford University Technical Report

Figure 14.   Acclimation to toxicity of individual cations (26)

**Table VIII.   Upper Limit of Cation Concentration in
Anaerobic Waste Treatment**

Cation	Slug Feed		Daily Feed	
	Single Cation, M	Antagonists Present, M	Single Cation, M	Antagonists Present, M
Sodium	0.2	0.3–0.35	0.3	70.35
Potassium	0.09	0.15–0.2	0.13	0.35
Calcium	0.07	0.125–0.15	0.15	0.20
Magnesium	0.05	0.125	0.065	0.14
Ammonium	0.1	0.25	Not measured	Not measured

greater than 90% of the control, a non-toxic situation is considered to exist.)

To rectify this situation, Chin, Kugelman, and Molof (25) studied cation effects in chemostat type reactors at various biological solids retention times. Chemostat reactors were used because the data from them are analyzed easily for fundamental kinetic information. The growth medium used was identical to that used by Kugelman and McCarty except for the substitution of New York City tap water for Cambridge tap water. Only two cations were investigated here, sodium and potassium, because of the large number of units and long time required to obtain data for just two cations.

Analysis of the data for fundamental kinetic information was based on the Monod model. This model has been used by many successfully to describe the kinetics of biological waste treatment systems (26). The fundamental equation of this model is:

$$u = \frac{k_m S}{k_s + S} \tag{1}$$

where u = specific growth rate, mg/(mg-day)

S = substrate concentration, mg/liter

k_m = maximum specific growth rate, mg/(mg-day)

k_s = substrate concentration when $u = k_m/2$, mg/liter

This equation can be restated in terms of substrate utilization rate as:

$$u_s = \frac{kS}{k_s + s} = \frac{k_m S}{Y_o(k_s + s)} \tag{2}$$

where u_s = specific substrate utilization rate, mg/(mg-day)

k = maximum specific substrate utilization rate, mg/(mg-day)

Y_o = yield constant (dimensionless)

A mass balance for organisms and substrate around a chemostat combined with these kinetic relationships yields the following two equations:

$$S = \frac{k_s \left[1 + k_d\,(SRT)\right]}{(SRT)\,k_m - \left[1 + k_d\,(SRT)\right]} \tag{3}$$

$$M = \frac{(S_o - S)\,Y_o}{\left[1 + k_d\,(SRT)\right]} \tag{4}$$

where S = effluent substrate concentration, mg/liter

S_o = influent substrate concentration, mg/liter

SRT = biological solids retention time, mg/liter

M = concentration of biological solids in the reactor, mg/liter

k_d = specific decay rate/, mg(mg-day)

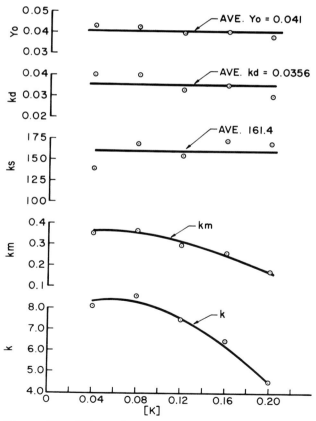

Figure 15.   Effect of potassium on kinetics of acetate degradation

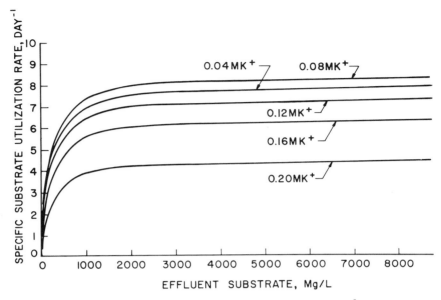

Figure 16.   Effect of potassium and acetate on acetate utilization rate

These four relationships contain five kinetic constants. If these are known along with SRT and S_o, the efficiency of operation of a waste treatment unit can be predicted. This kinetic model is for growth limited by an organic substrate and assumes that all other growth factors are optimum. In this study the cation concentration was certainly not optimum nor was it constant. The above kinetic model however could be used because what was desired was evaluation of the effect of changes in cation concentration on the "kinetic constants."

The effect of potassium on the kinetic constants is illustrated in Figure 15. Only the maximum specific growth rate, k_m, and the maximum specific substrate utilization rate, k, were affected by increases in potassium. These data were used to plot Figure 16, which indicates that the toxic effect of high concentrations of potassium is caused by a reduction in the rate at which each methane organism can process acetate.

Previous work had established sodium as an antagonist of potassium. Therefore, sodium was added to the units containing $0.2M$ potassium, and the effect on the kinetic constants was determined. Figure 17 shows that sodium reversed the adverse effect the high concentration of potassium had on k_m and k. Indeed, at the optimum sodium concentration of $0.03M$, these values were higher than in the system containing low potassium and no added sodium. This is similar to the situation of stimulation by the antagonist to a level of performance superior to the control observed by Kugelman and McCarty.

A similar set of studies was conducted for sodium toxicity. Figure 18 illustrates the effect of this cation on the kinetic constants. Here a different situation exists—while the maximum specific growth rate, k_m, is slightly affected, the major effects are a lowering of the yield constant Y_o and a significant increase in the specific organism decay rate k_d. Figure 19, which shows the specific substrate utilization rate as a function of substrate concentration and sodium concentration, again indicates a situation very different from that when the level of potassium was high. Both figures indicate that the adverse effect of sodium is not manifest as a reduced ability of each organism to process acetate. Rather the number of active organisms is lower than it should be because of the lower yield and increased rate of endogenous metabolism. The antagonistic action of

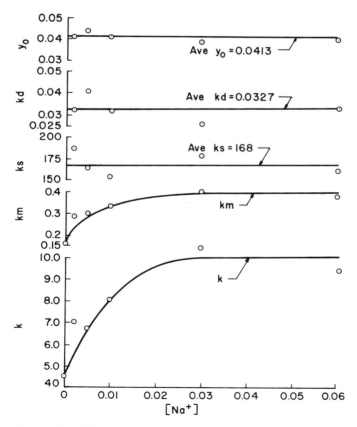

Figure 17.   Effect of sodium on kinetic constants in a system retarded by potassium

potassium to sodium toxicity is illustrated in terms of its effect on the kinetic constants in Figure 20. The addition of $0.03M$ potassium is sufficient to restore Y_o and k_d to the values they had when sodium was much lower than $0.35M$. These results confirm previous findings (24) that sodium adversely affects net synthesis in daily feed anaerobic waste treatment systems.

The greatest significance of this study is not the data generated, although they are significant data, but in the new approach it offers to handling toxicity data in the design of waste treatment systems. It is no longer necessary to report, for example, that sodium at a certain level is toxic or that it reduces the reaction rate to a certain fraction of the control. The correct statement is that the value of each of the kinetic constants is "such and such" at this level of sodium. The values of the kinetic constants can then be used to construct design charts such as that of Figure 21. This figure depicts substrate removal efficiency as a function of SRT and potassium concentration. If the designer is confronted

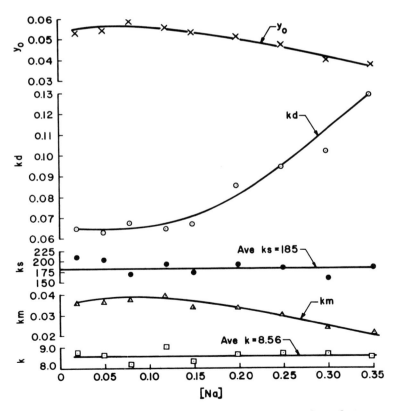

Figure 18.   Effect of sodium on kinetics of acetate degradation

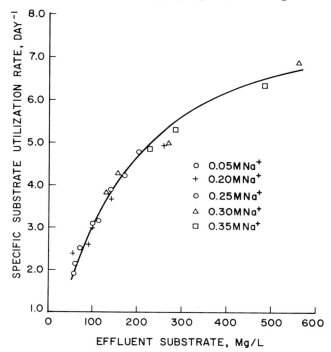

Figure 19.   Effect of sodium and acetate on acetate utilization rate

with a waste containing $0.2M$ potassium, this design chart indicates he must provide an *SRT* of at least 11 days to achieve a waste reduction rate of 90%. Alternately, the designer could determine from a chart such as Figure 22 that he could obtain the same efficiency at an *SRT* of six days by adding $0.01M$ sodium to the waste.

If this method is used in reporting data, it will no longer be necessary to use non-quantitative terms such as toxicity, antagonism, synergism, stimulation, and etc. Because of the absolute nature of this method of reporting results and its freedom from ambiguity, all future studies of the effect of metabolites on metabolism should utilize this reporting method.

Miscellaneous Toxic Materials

Many other groups of substances have been reported to exert a toxic effect on anaerobic waste treatment systems. However, the experimental data for most of these are quite meager. In this section some of the available data on these substances are given.

McCarty and McKinney reported (*18*) that pH played a significant role in the toxicity of ammonium ion. They deduced that free ammonia produced from ammonium as the pH increased above 7 was the actual toxic agent. It was found that if the free ammonia concentration exceeded 150 mg/liter, severe toxicity resulted. It is easy to avoid ammonia toxicity by adding HCl to keep the pH close to 7.0.

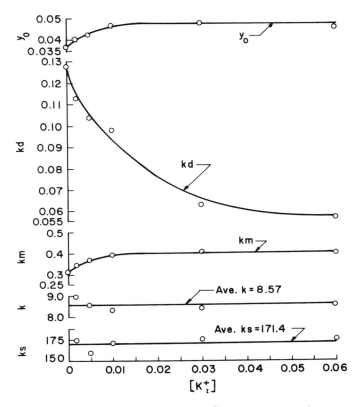

Figure 20. *Effect of potassium on kinetic constants in a system retarded by sodium*

In an earlier section it was reported that sulfides were excellent for preventing heavy metal toxicity. McCarty *et al.* (*21*) reported however, that soluble sulfide is toxic if the concentration exceeds 200 mg/liter. The soluble sulfide concentration in a digester is a function of the incoming sources of sulfur, the pH, the rate of gas production, and the availability of heavy metals to act as precipitants. Excess sulfide can be removed by adding iron to precipitate FeS.

Long chain fatty acids such as palmitic, stearic, and oleic acids can exert a toxic effect in anaerobic digestion if they are in solution. Highly

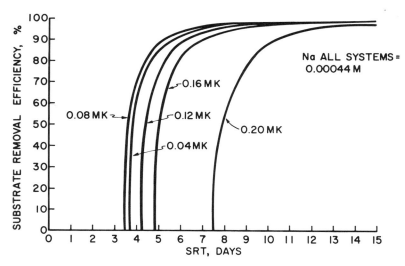

Figure 21. *Computed substrate removal efficiency at different potassium concentrations with no sodium supplement*

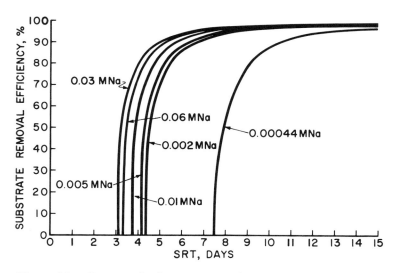

Figure 22. *Computed substrate removal efficiency at 0.2 MK and different sodium concentrations*

insoluble calcium salts of these acids can be formed to prevent toxicity (9).

Slug additions of 50 mg/liter of nitrate have been reported as harmful to methane bacteria (27). If however, facilitative bacteria are present, they will rapidly reduce the nitrate and prevent toxicity (9).

Among other substances reported as toxic to methane bacteria are phenols (28), methane analogs such as chloroform (29), and ABS. As noted above little significant data exists on the toxicity of these substances.

Conclusions

(1) Much of the published data on toxicity in anaerobic waste treatment systems is erroneous and misleading because of inadequate experimental techniques.

(2) The major experimental inadequacies include improper characterization of the control, neglect of antagonism, synergism, acclimation, and complexing reactions.

(3) The toxicity of heavy metals can be eliminated completely by precipitation as sulfides. Ferrous sulfate is the favored source of sulfide. Sodium sulfide can be used, but care must be exercised to avoid sulfide toxicity.

(4) Volatile organic acids are not toxic to methane bacteria at concentrations up to 6000 mg/liter. Propionic acid is slightly toxic at this level to acid-forming bacteria. Thus, pH control with alkaline substances during unbalanced digestion is a valid procedure. However, care must be exercised that the cation of the alkaline material does not produce a toxic condition.

(5) Toxicity of light metal cations is significantly affected by the presence or absence of extremely low concentrations of antagonists and synergists. When dealing with cation toxicity situations, it is essential that a complete ionic analysis of the environment be made so that valid decisions on ionic alterations can be made.

(6) Antagonism by cations is associated with the optimum nutritional requirement for each cation. The optimum ionic environment is approximately $0.01M$ for monovalent cations and $0.005M$ for divalent cations.

(7) Toxicity data and its associated phenomena are usually presented with reference to a control unit. A more valid procedure especially for use by the design engineer would be presentation on an absolute basis. This can be achieved by determining the variation of kinetic parameters with change in concentration for the substances under study.

(8) With light metal cations it is valid to use the yield constant (Y_o), maximum specific growth rate (k_m), specific decay rate (k_d), and half maximum velocity constant (k_s), as defined for the Monod model as the kinetic parameters to categorize toxicity data. It is suggested that this procedure be applied to all substances.

(9) A great deal of additional work is required to provide the design engineer with the information he needs to deal rationally with toxicity problems in anaerobic waste treatment.

Literature Cited

(1) Niles, A. H., Frock, J. E., "Studies on Digestion End Liquor from Production of Monosodium Glutamate," *Proc. Ind. Waste Conference, 7th, Purdue University, 1952,* 269.

(2) Oliver, G. E., Dunsten, G. H., "Anaerobic Digestion of Pea Blancher Wastes," *Sewage Ind. Wastes* (1955) **27**, 1171.

(3) Buswell, A. M., Pagano, J. F., Sollo, F. W., "Effect of Sodium Sulfate and Sodium Chloride on Methane Fermentation," *Ind. Eng. Chem.* (1949) **41**, 596.

(4) Hotchkiss, M., "Studies on Salt Action. VI. The Stimulating and Inhibitive Effect of Certain Cations on Bacterial Growth," *J. Bacteriol.* (1923) **8**, 141.

(5) Falk, I. S., "The Role of Certain Ions in Bacterial Physiology, A Review," *Abstr. Bacteriol.* (1923) **7**, 33–50, 87–105, 133–147.

(6) Toaca, P., Wilcox, E., Finland, M., "Effect of pH of Medium and Site of Inoculum on Activity of Antibiotics Against Group D Streptococcus," *Appl. Microbiol.* (1970) **19**, 629.

(7) Barth, E. F., Moore, W. A., McDermott, G. N., "Interaction of Heavy Metals an Biological Sewage Treatment Processes," U. S. Department of Health, Education and Welfare, May 1965.

(8) Masselli, J. W., Masselli, N. W., Burford, G., "The Occurrence of Copper in Water, Sewage and Sludge and its Effect on Sludge Digestion," New England Interstate Water Pollution Control Commission, Boston, Mass. (June 1961).

(9) McCarty, P. L., Kugelman, I. J., Lawrence, A. W., "Ion Effects in Anaerobic Digestion," *Tech. Rept.* No. 33, Department of Civil Engineering, Stanford University, Stanford, Calif. (1964).

(10) Lawrence, A. W., McCarty, P. L., "The Role of Sulfide in Preventing Heavy Metal Toxicity in Anaerobic Treatment," *J. Water Pollut. Contr. Fed.* (1965) **37**, 392–409.

(11) Barker, H. A., "Biological Formation of Methane," *Ind. Eng. Chem.* (1956) **48**, 1438.

(12) Cassel, E. A., Sawyer, C. N., "A Method of Starting Up High Rate Digesters," *Sewage Ind. Wastes* (1959) **31**, 123.

(13) Fair, G. M., Carlson, C. L., "Sludge Digestion—Reaction and Control," *J. Boston Soc. Civil Eng.* (1927) **14**, 82.

(14) Heukelekian, H., "Volatile Acids in Digesting Sludge," *Ind. Eng. Chem.* (1928) **20**, 752.

(15) Buswell, A. M., "Important Considerations in Sludge Digestion. II. Microbiology and Theory of Anaerobic Digestion," *Sewage Works J.* (1947) **11**, 28.

(16) Schlenz, H. E., "Important Considerations in Sludge Digestion. I. Practical Aspects," *Sewage Works J.* (1947) **19**, 19.

(17) McCarty, P. L., McKinney, R. E., "Volatile Acid Toxicity in Anaerobic Digestion," *J. Water Pollut. Contr. Fed.* (1961) **33**, 223.

(18) McCarty, P. L., McKinney, R. E., "Salt Toxicity in Anaerobic Digestion," *J. Water Pollut. Contr. Fed.* (1961) **33**, 399.

(19) Jeris, J. S., McCarty, P. L., "The Biochemistry of Methane Fermentation Using C^{14} Tracers," *Proc. Ann. Purdue Ind. Waste Conf., 17th, 1962.*

(20) Buswell, A. M., Morgan, G. B., "Paper Chromatographic Method for Volatile Acid Determination. Part III. Toxicity of Propionic Acid," *Tech. Paper* No. **239**, Florida Engineering and Industrial Experiment Station (1962) **16**, 10.

(21) Rudolphs, W., Zeller, P. J. A., "Odors and Sewage Sludge Digestion, I. Effect of Sea Water on Hydrogen Sulfide Production," *Ind. Eng. Chem.* (1928) **20**, 48.

(22) Winslow, C. E. A., Haywood, E. T., "The Specific Potency of Certain Cations with Reference to Their Effect of Bacterial Visibility," *J. Bacteriol.* (1931) **22**, 49.

(23) Kugelman, I. J., McCarty, P. L., "Cation Toxicity and Stimulation in Anaerobic Waste Treatment. I. Slug Feed Studies," *J. Water Pollut. Contr. Fed.* (1965) **37**, 97–115.

(24) Kugelman, I. J., McCarty, P. L., "Cation Toxicity and Stimulation in Anaerobic Waste Treatment. II. Daily Feed Studies," *Proc. Ind. Waste Treatment Conf., 19th, Purdue University, 1965,* 667–686.

(25) Chin, K. K., Kugelman, I. J., Molof, A. H., "The Effect of Monovalent Cations on the Methane Phase in Continuous Digestion Systems," *J. Water Pollut. Contr. Fed.,* in press.

(26) Lawrence, A. W., McCarty, P. L., "Kinetics of Methane Fermentation in Anaerobic Waste Treatment," *Tech. Rept.* No. **75**, Department of Civil Engineering, Stanford University, Stanford, Calif. (1967).

(27) Buswell, A. M., Pagano, J. F., "Reduction and Oxidation of Nitrogen Compounds in Polluted Streams," *Sewage Ind. Wastes* (1952) **24**, 897.

(28) Jeris, J., private communication.

(29) Bauchop, T., "Inhibition of Rumen Methanogenesis by Methane Analogues," *J. Bacteriol.* (1967) **94**, 171.

(30) Rudgel, H. T., "Bottle Experiments as a Guide in Operation of Digesters Receiving Copper Sludge Mixtures," *Sewage Works J.* (1941) **13**, 1248.

(31) Rudgel, H. T., "Effects of Copper Bearing Wastes on Sludge Digestion," *Sewage Works J.* (1946) **18**, 1130.

(32) Barnes, G. E., Braidech, M. W., "Treating Pickling Liquors for Removal of Toxic Metals," *Eng. News Record* (1942) **129**, 496.

(33) Wischmeyer, W. J., Chapman, J. T., "A Study of the Effect of Nickel on Sludge Digestion," *Sewage Works J.* (1947) **19**, 790–5.

(34) Rudolphs, W., Zeller, P. J. A., "Effect of Sulfate Salts on Hydrogen Sulfide Production in Sludge Digestion," *Sewage Works J.* (1932) **4**, 77.

(35) McDermott, G. N., Barth, E. F., Salotto, V., Ettinger, M. B., "Zinc in Relation to Activated Sludge and Anaerobic Digestion Processes," *Proc. Purdue Industrial Waste Conference, 17th, 1963, XLVII,* 461–475.

(36) Pagano, J. F., Teweles, R., Buswell, A. M., "The Effect of Chromium on the Methane Fermentation of Acetic Acid," *Sewage Ind. Wastes* (1950) **22**, 336–345.

RECEIVED August 3, 1970.

6

Energetics and Kinetics of Anaerobic Treatment

PERRY L. McCARTY

Department of Civil Engineering, Stanford University, Stanford, Calif. 94305

A steady-state model which considers the anaerobic process as a series of bacterially mediated reactions is presented. The organism yield is considered to be a function of reaction energetics. The analysis allows computation of the concentration of all components and organism concentrations as a function of the cellular retention time. The results from the model are qualitatively in agreement with experimental results. Closer agreement requires a better knowledge of the stoichiometry of individual reactions. The model can be useful in ascertaining what this stoichiometry may be and can be helpful in seeking a better understanding of the anaerobic treatment process.

The anaerobic process is an interesting one from a thermodynamic point of view because the energy change per mole of substrate fermented is so small, often no more than a few kilocalories per mole. In addition, many organic compounds are fermented to methane gas in a step process by different groups of organisms operating in series. Thus, the relatively little energy yield from the over-all conversion of complex organics to methane gas must be divided into even smaller packets and distributed among the different bacteria involved.

Many of the important bacteria which mediate the over-all methane fermentation have not been isolated and questions about pathways are still unanswered. The use of thermodynamics may help in suggesting which of the many possible reactions are likely, as well as helping to explain some of the observations about the process. The thermodynamic and kinetic analysis presented in this paper is applied to steady-state reactor operation.

Steady-State Model Development

To illustrate the influence of reaction energetics on cellular yield and other characteristics of the methane fermentation, and to simplify the analysis, a few assumptions will be made.

(1) Reactions occur in a continuous flow stirred anaerobic reactor without recycle and operating under steady-state conditions.

(2) No inhibitors to biological growth are present within the reactor.

(3) Decay rate of microorganisms is zero.

(4) Microorganism concentration in the influent stream to the reactor is zero but is finite for all required species within the reactor itself.

(5) Reactions follow Michaelis-Menten type kinetics.

(6) A given waste component can be rate limiting for one reaction only.

(7) A given species of bacteria mediates only a single reaction.

(8) Cellular yield of organisms is a function of reaction energetics.

(9) Rate of electron transfer in energy yielding reactions is a temperature-dependent constant for all microorganisms.

Several of these assumptions do not have general validity for all biological reactions but are not felt to limit severely the analysis in the present context. Assumptions 6 and 7 are certainly not of general usefulness but are quite reasonable for methane fermentations. A unique characteristic of these organisms is that their ability to mediate reactions is very limited, and several different species are required to bring about the conversion of a complex substrate to methane gas. The last two assumptions have been found to be reasonably valid for microorganisms which tend to dominate in natural environments. These assumptions are basic to the present analysis, and the limitations which result are governed primarily by limitations in these two assumptions.

With the above constraints, a continuous flow stirred anaerobic reactor as illustrated in Figure 1 will be considered to operate under steady-state conditions $(dQ/dt = dn_i^o/dt = dn_i^g/dt = dn_i^a/dt = 0)$, where Q is the flow rate in liters per day and n_i^o, n_i^g, and n_i^a represent the number of moles of component A_i per unit time entering the reactor, and leaving the reactor in the gas and aqueous phases, respectively. The system contains m components A_i $(i = 1, 2, \ldots . m)$ interconnected by means of n biologically mediated reactions B_k $(k = 1, 2, \ldots . n)$. There are more components than reactions so that n is less than m. S_i is the concentration of component A_i in the reactor in moles per liter. The reactor contains finite concentrations of all components.

The equation for reaction B_k can be written in the following generalized way:

$$0 = \sum_{i=1}^{m} v_{ik} A_i \qquad (1)$$

The stoichiometric coefficient v_{ik} for component A_i in reaction B_k is positive for products and negative for reactants and equals zero for components that do not take part in the reaction.

From the assumptions previously listed, the rate of reaction B_k depends only upon the functional characteristics of the organisms mediating the reaction and upon the activities of the reactants and products for the reaction. The microorganisms mediating reaction B_k are a different species from those mediating any other reaction.

The rate of reaction B_k is r_k and is considered to be adequately described by Michaelis-Menten type kinetics in which:

$$r_k = \frac{k_k X_k S_k}{K_k + S_k} \tag{2}$$

X_k is the concentration of microorganisms mediating the reaction. Based upon Assumption 7, one component is rate limiting for each reaction. By special ordering of reactions and components, it can be prescribed that component A_k is the rate limiting component for reaction B_k. Thus, S_k becomes the concentration of component A_k which is rate limiting for reaction B_k. K_k and k_k are temperature-dependent coefficients for the reaction.

A component enters the reactor in aqueous solution and may leave the reactor either with the aqueous phase or as a gas. A materials balance for component A_i for the system will yield the following:

$$n_i^o - n_i^a - n_i^g = \sum_{k=1}^{n} (-v_{ik}) r_k V \tag{3}$$

V is the reactor volume in liters. The designation X is used when referring to microorganisms. The stoichiometric coefficient v_{xk} then is used as the stoichiometric coefficient for organism growth in reaction B_k. For the steady-state case where X_k^o, the influent concentration of organisms mediating reaction B_k, is zero, the rate of increase in organism mass resulting from substrate utilization in the reactor ($v_{xk} r_k V$) must equal the rate of loss from the reactor ($Q X_k$):

$$v_{xk} r_k V = X_k Q \tag{4}$$

or

$$X_k = v_{xk} r_k \theta_c$$

θ_c is defined as the cell retention time and for the continuous flow reactor illustrated in Figure 1 equals the theoretical hydraulic residence time (V/Q). Combining Equations 2 and 4 results in an equation relating the concentration of component A_k to θ_c:

$$S_k = \frac{K_k}{v_{xk} k_k \theta_c - 1}, r_k > 0 \tag{5}$$

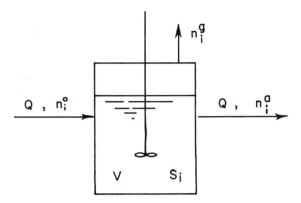

*Figure 1. Schematic of continuous flow stirred
anaerobic reactor*

Equation 5 is meaningful only if the reaction rate r_k is greater than
zero. The concentration of S_k in the general case can be found from a
mass balance around the reactor using Equation 3. Converting from
moles to concentration terms, Equation 3 will yield the following for the
concentration of component A_k:

$$S_k = \theta_c \left[\frac{n_k{}^o - n_k{}^g}{V} - \sum_{j=1}^{n} (-v_{kj}r_j) \right] \qquad (6)$$

If S_k as determined from Equation 6 is less than S_k as determined by
Equation 5, then r_k equals zero as does X_k the concentration of microorganisms mediating the reaction. The significance of this statement is that
when S_k drops below the critical value as given by Equation 5, the rate
of formation of microorganisms mediating the reaction is less than the
rate at which they are removed from the reactor so that in the steady
state, X_k and hence r_k equal zero.

For non-gaseous components, $n_i{}^g$ will equal zero. For gaseous components, there will be a distribution of the component between $n_i{}^a$ and
$n_i{}^g$ depending upon the solubility of the gas and the relative total rates
of gas production and detention time of fluid within the reactor. When
making a mass balance it will be assumed, however, that $n_i{}^a$ for gaseous
components equals zero. When component A_i is present as a gas, its partial pressure in the reactor in atmospheres is designated p_i and can be
determined by:

$$p_i = \frac{n_i{}^g}{\sum\limits_{j=1}^{m} n_j{}^g} \qquad (7)$$

For a given waste stream, series of reactions B_k, components A_i, and cell retention time θ_c, the concentration of all components in the reactor effluent stream and gas discharge can be determined. This requires that the stoichiometric coefficients v_{ik} be known, including the value v_{xk} for organism synthesis. Also required are the coefficients k_k and K_k for each reaction. The method used for selection of appropriate values are discussed in the following.

Evaluation of Stoichiometric Coefficients

Reaction equations with proper stoichiometric coefficients can be devised for over-all biological reactions by first evaluating the reactions for energy and synthesis separately and then adding them together. Heterotrophic and chemosynthetic autotrophic microorganisms obtain energy for growth by mediating oxidation-reduction reactions. Such reactions involve a flow of electrons from electron donors to electron acceptors. A partial list of half-equations which can be combined in various ways to yield equations for oxidation-reduction reactions of interest in methane fermentation are listed in Table I. The last equation contains an empirical molecular formula for bacterial cells ($C_5H_7O_2N$) and is useful in constructing synthesis equations.

Table II contains representative over-all reactions for methane fermentations of acetate and hydrogen, respectively. Acetate fermentation is mediated by heterotrophic organisms which use acetate as a carbon source for synthesis as well as energy. For reaction B_k, E_k represents the electron equivalents of the electron donor converted for energy per electron equivalent of cells synthesized. Hydrogen fermentation, by contrast, is mediated by autotrophic organisms which use carbon dioxide, acetate, or some other carbon source for cell synthesis. In the equation shown in Table II acetate was assumed to be the carbon source used as found for this fermentation by Bryant (*1*). E_k has the same definition as for heterotrophic growth.

The stoichiometric coefficient v_{ik} for each of the over-all reactions in Table II equals the sum of the coefficients for the energy reaction $v_{ik}{}^e$ and the synthesis $v_{ik}{}^s$:

$$v_{ik} = v_{ik}{}^e + v_{ik}{}^s \tag{8}$$

While the values for E_k were assumed for the examples listed in Table II, they can be approximated from considerations of energy released by the energy reactions, the energy required for synthesis, and the efficiency by which energy is transferred in these reactions (*2*). The Gibbs free energy

Table I.   Free Energies for

Half-

---

(1) $\frac{1}{2}$ H$_2$O

(2) $\frac{1}{8}$ CH$_4$ + $\frac{1}{4}$ H$_2$O

(3) $\frac{1}{8}$ CH$_3$COO$^-$ + $\frac{3}{8}$ H$_2$O

(4) $\frac{1}{92}$ CH$_3$(CH$_2$)$_{14}$COO$^-$ + $\frac{31}{92}$ H$_2$O

(5) $\frac{1}{14}$ CH$_3$CH$_2$COO$^-$ + $\frac{5}{14}$ H$_2$O

(6) $\frac{1}{20}$ CH$_3$CH$_2$CH$_2$COO$^-$ + $\frac{7}{20}$ H$_2$O

(7) $\frac{1}{12}$ CH$_3$CH$_2$OH + $\frac{1}{4}$ H$_2$O

(8) $\frac{1}{12}$ CH$_3$CHNH$_2$COOH + $\frac{5}{12}$ H$_2$O

(9) $\frac{1}{10}$ CH$_3$COCOO$^-$ + $\frac{2}{5}$ H$_2$O

(10) $\frac{1}{2}$ H$_2$

(11) $\frac{1}{24}$ C$_6$H$_{12}$O$_6$ + $\frac{1}{4}$ H$_2$O

(12) $\frac{1}{2}$ HCOO$^-$ + $\frac{1}{2}$ H$_2$O

(13) Bacterial Cells:
$\frac{1}{20}$ C$_5$H$_7$O$_2$N + $\frac{9}{20}$ H$_2$O

Various Half-Reactions

Reaction	ΔG^o, kcal per mole electrons
$= \frac{1}{4} O_2 + H^+ + e^-$	28.345
$= \frac{1}{8} CO_2 + H^+ + e^-$	3.907
$= \frac{1}{8} CO_2 + \frac{1}{8} HCO_3^- + H^+ + e^-$	3.061
$= \frac{15}{92} CO_2 + \frac{1}{92} HCO_3^- + H^+ + e^-$	3.013
$= \frac{1}{7} CO_2 + \frac{1}{14} HCO_3^- + H^+ + e^-$	3.006
$= \frac{1}{20} HCO_3^- + \frac{3}{20} CO_2 + H^+ + e^-$	2.917
$= \frac{1}{6} CO_2 + H^+ + e^-$	2.078
$= \frac{1}{6} CO_2 + \frac{1}{12} HCO_3^- + \frac{1}{12} NH_4^+ + H^+ + e^-$	2.031
$= \frac{1}{5} CO_2 + \frac{1}{10} HCO_3^- + H^+ + e^-$	1.125
$= H^+ + e^-$	0.000
$= \frac{1}{4} CO_2 + H^+ + e^-$	−0.350
$= \frac{1}{2} HCO_3^- + H^+ + e^-$	−1.810
$= \frac{1}{5} CO_2 + \frac{1}{20} HCO_3^- + \frac{1}{20} NH_4^+ + H^+ + e^-$	

change (ΔG_k) for the energy portion of reaction B_k can be written as:

$$\Delta G_k = \Delta G_k{}^o + RT \sum_{i=1}^{m} v_{ik}{}^e \ln a_i \qquad (9)$$

where $\Delta G_k{}^o$ is the standard free energy for reaction B_k as determined from the values listed in Table I and as illustrated in Table II, and a_i is the activity of component A_i. The activity for non-gaseous components is approximated by S_i, and for gaseous components by p_i.

McCarty (2) presented the following equation for the estimation of E_k:

$$E_k = - \frac{\Delta G_{pk}/e^m + \Delta G_{ck} + \Delta G_{nk}/e}{e\Delta G_k} \qquad (10)$$

Table II.   Energy and Synthesis Reactions for

Substrate	Reaction Type	Half-Reactions from Table I	Assumed Value of E_k
Acetate	Energy	(3)-(2)	
	Synthesis	(3)-(13)	20
	Over-all		20
Hydrogen	Energy	(10)-(2)	
	Synthesis	(3)-(13)	5
	Over-all		5

The terms in the numerator represent the quantity of energy required to synthesize an electron equivalent of cells. The term ΔG_{pk} represents the quantity of energy required to convert the carbon source used for synthesis of cells mediating reaction B_k into an intermediate (assumed to be pyruvate for energy calculations). This is obtained by subtracting ΔG^0 for the pyruvate half reaction (Equation 9, Table I) from ΔG^0 for the half-reaction involving the cell carbon source. ΔG_{ck} represents the energy for conversion of intermediate into cells and was estimated to equal 7.5 kcal/electron mole of cells. ΔG_{nk} represents the energy required to reduce the nitrogen source used for cell synthesis into ammonia. ΔG_{nk} equals zero if the nitrogen source is ammonia, the usual nitrogen source in methane fermentations.

Methane Fermentations of Acetate and Hydrogen

Reaction	ΔG_k^o kcal per electron equivalent
$\frac{1}{8} CH_3COO^- + \frac{1}{8} H_2O = \frac{1}{8} HCO_3^- + \frac{1}{8} CH_4$	-0.846
$\frac{1}{20}\left[\frac{1}{8} CH_3COO^- + \frac{3}{40} CO_2 + \frac{1}{20} NH_4^+ =\right.$	
$\left.\frac{1}{20} C_5H_7O_2N + \frac{3}{40} HCO_3^- + \frac{13}{40} H_2O\right]$	

$0.131\ CH_3COO^- + 0.0025\ NH_4^+ + 0.0038\ CO_2 +$
$\quad 0.109\ H_2O = 0.0025\ C_5H_7O_2N + 0.125\ CH_4 +$
$\quad 0.129\ HCO_3^-$

Reaction	ΔG_k^o
$\frac{1}{2} H_2 + \frac{1}{8} CO_2 = \frac{1}{8} CH_4 + \frac{1}{4} H_2O$	-3.91
$\frac{1}{5}\left[\frac{1}{8} CH_3COO^- + \frac{3}{40} CO_2 + \frac{1}{20} NH_4^+ =\right.$	
$\left.\frac{1}{20} C_5H_7O_2N + \frac{3}{40} HCO_3^- + \frac{13}{40} H_2O\right]$	

$0.5\ H_2 + 0.025\ CH_3COO^- + 0.14\ CO_2 + 0.01\ NH_4^+ =$
$\quad 0.01\ C_5H_7O_2N + 0.125\ CH_4 + 0.015\ HCO_3^- +$
$\quad 0.315\ H_2O$

The denominator of Equation 10 represents the energy released and available for the synthesis reaction per electron equivalent of change for the energy reaction. The value ΔG_k is a function of the activities of reactants and products and is determined by use of Equation 9. The value of e represents the efficiency of energy conversion, and for autotrophic and heterotrophic bacteria growing under suitable environmental conditions generally varies from about 0.4 to 0.8 with an average of about 0.6. The value for m is $+1$ for $\Delta G_p > 0$ and -1 for $\Delta G_p < 0$.

Evaluation of Kinetic Coefficients

The two temperature dependent coefficients k_k and K_k in Equation 2 must be evaluated by some means to evaluate the significance of thermodynamics on the over-all methane fermentation process. The maximum electron transfer rate, k_k, is quite similar for a variety of heterotrophic and autotrophic bacteria (2), equaling 1 to 2 electron moles/gram dry weight of bacteria per day at 25°C. Using the Arrhenius equation,

$$k_k = D \exp(-H_a/RT) \tag{11}$$

the value for H_a within a temperature range of 10° to 40°C varies between about 9 and 18 kcal/mole, R equals 1.99 kcal mole^{-1} degree^{-1}, and T is temperature in degrees Kelvin.

The value for K_k, the saturation coefficient or growth-limiting substrate concentration at which organism growth rate is one-half its maximum value, varies considerably with the reaction, the species mediating the reaction, and environmental conditions. In anaerobic methane fermentations, K_k tends to be much larger, at least with acetate and propionate, than for aerobic oxidations. Values of 10^{-3} moles/liter seem appropriate for the former and 10^{-4} for the latter. K_k also tends to increase with decrease in temperature.

Model Application to a Complex Anaerobic System

A chain of reactions for the biological conversion of carbohydrates, proteins and fats into various intermediates and finally into methane is illustrated in Figure 2. This chain represents a greatly simplified but realistic separation of methane fermentation into the various steps of most importance. Table III lists the energy and synthesis reactions for each step, including the stoichiometric coefficients for each component in each reaction, and also lists the limiting component for each reaction. Carbohydrate is assumed to be fermented to hydrogen, acetate, propionate, and butyrate in proportion to the average stoichiometric quantities observed to result from cellulose fermentation in the rumen of animals

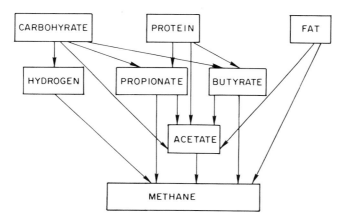

Figure 2.   Chain of reactions for anaerobic biological conversion of carbohydrates, proteins, and fats into methane

(3). Similar information was not available for protein fermentation, and so this fermentation was assumed with little factual basis to result in the production of approximately equal molar concentrations of acetate, propionate, and butyrate. The fat in municipal wastewater consists primarily of even carbon fatty acids which have been verified to be fermented by beta oxidation to methane and acetate (4), resulting quite closely in this case to the stoichiometry given by the listed reaction.

The fermentation of carbohydrates and protein are each no doubt mediated by more than one species of bacteria, and so the assumption that this occurs in a single step is a significant simplification. However, the fermentation of hydrogen and the various fatty acids by independent species of methane producing bacteria appears to be quite realistic.

A waste with characteristics as listed in Table IV has been assumed. The assumed composition is similar to that for a municipal primary treatment waste sludge for which data on methane fermentation was available for comparison. The concentrations are expressed in millielectron equivalents since the materials balances and calculations were made on this basis. One millielectron equivalent per liter is equivalent to 8 mg/liter of chemical oxygen demand (COD), a more commonly used measure of waste strength.

With the values listed in Table IV, the reaction equations listed in Table III, and with the use of Equations 1 through 10, the effect of varying θ_c on the reactor performance and the resulting distribution of the various components involved was determined.

The results of the analysis are illustrated in Figure 3. Protein and carbohydrate fermentation occur extensively at θ_c values of less than one day, while fermentation of the fatty acids formed occurs only at θ_c values

Table III.   Reactions Used to Evaluate Methane

Limiting Substrate A_k	*Reaction*
Carbohydrate	Energy $0.167\ C_6H_{12}O_6 + 0.286\ HCO_3^- =$
	Synthesis: $\dfrac{1}{24}\ C_6H_{12}O_6 + \dfrac{1}{20}\ HCO_3^- + \dfrac{1}{20}\ NH_4^+ =$
Protein	Energy: $0.33\ C_4H_6ON + 0.073\ HCO_3^- + 0.64\ H_2O =$
	Synthesis: $\dfrac{1}{17}\ C_4H_6ON + \dfrac{7}{340}\ H_2O + \dfrac{2}{85}\ CO_2 =$
Fat	Energy: $\dfrac{1}{28}\ C_{15}H_{31}COO^- + \dfrac{1}{4}\ HCO_3^- =$
	Synthesis: $\dfrac{1}{92}\ C_{15}H_{31}COO^- + \dfrac{9}{230}\ HCO_3^- + \dfrac{1}{20}\ NH_4^+ =$
Hydrogen	Energy: $\dfrac{1}{2}\ H_2 + \dfrac{1}{8}\ CO_2 =$
	Synthesis: $\dfrac{1}{8}\ CH_3COO^- + \dfrac{3}{40}\ CO_2 + \dfrac{1}{20}\ NH_4^+ =$
Propionate	Energy: $\dfrac{1}{6}\ CH_3CH_2COO^- + \dfrac{1}{12}\ H_2O =$
	Synthesis: $\dfrac{1}{14}\ CH_3CH_2COO^- + \dfrac{1}{20}\ NH_4^+ + \dfrac{2}{35}\ CO_2 =$
Butyrate	Energy: $\dfrac{1}{4}\ CH_3CH_2CH_2COO^- + \dfrac{1}{4}\ HCO_3^- =$
	Synthesis: $\dfrac{1}{20}\ CH_3CH_2CH_2COO^- + \dfrac{1}{20}\ CO_2 + \dfrac{1}{20}\ NH_4^+ =$

Fermentation of Carbohydrate, Protein, and Fat Mixture

	kcal/electron equivalent	
	$\Delta G_k{}^\circ$	ΔG_{p_k}
$0.177\ CH_3COO^- + 0.063\ CH_3CH_2COO^- +$ $0.0466\ CH_3CH_2CH_2COO^- + 0.555\ CO_2 +$ $0.385\ H_2 + 0.172\ H_2O$	-10.86	
$\frac{1}{20}\ C_5H_7O_2N + \frac{1}{20}\ CO_2 + \frac{1}{5}\ H_2O$		-1.48
$0.33\ NH_4^+ + 0.14\ CH_3COO^- + 0.13\ CH_3CH_2COO^- +$ $0.133\ CH_3CH_2COO^- + 0.193\ CO_2$	-6.07	
$\frac{1}{20}\ C_5H_7O_2N + \frac{3}{340}\ NH_4^+ + \frac{3}{340}\ HCO_3^-$		0.85
$\frac{2}{7}\ CH_3COO^- + \frac{1}{8}\ CH_4 + \frac{1}{8}\ CO_2$	-1.00	
$\frac{1}{20}\ C_5H_7O_2N + \frac{13}{115}\ CO_2 + \frac{13}{115}\ H_2O$		1.89
$\frac{1}{8}\ CH_4 + \frac{1}{4}\ H_2O$	-3.91	
$\frac{1}{20}\ C_5H_7O_2N + \frac{3}{40}\ HCO_3^- + \frac{3}{70}\ H_2O$		1.94
$\frac{1}{6}\ CH_3COO^- + \frac{1}{8}\ CH_4 + \frac{1}{24}\ CO_2$	-0.97	
$\frac{1}{20}\ C_5H_7O_2N + \frac{3}{140}\ HCO_3^- + \frac{13}{140}\ H_2O$		1.88
$\frac{1}{2}\ CH_3COO^- + \frac{1}{8}\ CH_4 + \frac{1}{8}\ CO_2$	-1.57	
$\frac{1}{20}\ C_5H_7O_2N + \frac{1}{10}\ H_2O$		1.79

Table III.

Limiting Substrate A_k	Reaction

Acetate

Energy: $\frac{1}{8} CH_3COO^- + \frac{1}{8} H_2O =$

Synthesis: $\frac{1}{8} CH_3COO^- + \frac{3}{40} CO_2 + \frac{1}{20} NH_4^+ =$

greater than 3 days. Hydrogen fermentation occurs at a θ_c less than 2 days. Methane production is limited until fatty acid fermentation becomes significant.

Y_k is the yield of cells from reaction B_k expressed in electron equivalents per electron equivalent of substrate consumed and equals $[1/(1 + E_k)]$. Values for Y_k are illustrated in Figure 4 to show both the relative yields for the various substrates and the influence of θ_c. Carbohydrate, protein, and hydrogen fermentation produce the highest relative yields as a result of the greater energy released per electron equivalent fermented. This high yield is a major factor resulting in effective fermentation at low θ_c values. The greater synthesis per unit of substrate fermented allows these organisms to reproduce faster so that they remain in the reactor at lower θ_c than for the methane organisms fermenting acetate. The predicted effect of changing concentrations of different constituents within the reactor, which affects the energetics of the reaction and hence

Table IV.   Waste Characteristics and Model Coefficients Assumed for Analysis of Methane Fermentation of Complex Substrates

Influent Waste Characteristics	Quantity
Carbohydrate concentration	512 millielectron equiv/liter
Protein	340 millielectron equiv/liter
Fat	1110 millielectron equiv/liter
Acetate, propionate, butyrate	0

Coefficients	
k_k	1.0 electron equiv/gm-day
K_k	10^{-3} mole/liter[a]
e_k	0.6
ΔG_{ni}	0
$[HCO_3^-]$	10^{-2} mole/liter
Pressure	1 atm

[a] $C_6H_{12}O_6$, C_4H_6ON, and $C_{15}H_{31}COOH$ considered as molecular formula for carbohydrate, protein, and fat, respectively.

Continued

<table>
<tr><td></td><td colspan="2">*kcal/electron equivalent*</td></tr>
<tr><td></td><td>ΔG_k°</td><td>ΔG_{p_k}</td></tr>
<tr><td>$\frac{1}{8} CH_4 + \frac{1}{8} HCO_3^-$</td><td>$-0.85$</td><td></td></tr>
<tr><td>$\frac{1}{20} C_5H_7O_2N + \frac{3}{40} HCO_3^- + \frac{3}{40} H_2O$</td><td></td><td>$1.936$</td></tr>
</table>

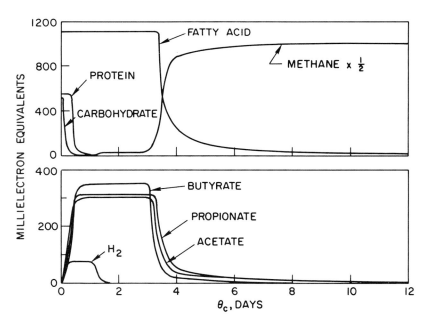

Figure 3.   Computed concentration of various components as a function of cell retention time

cell yield coefficients, is also seen from the figure. The effect is different with different substrates and is a result of their varying interdependencies.

Figure 5 is a summary of the total cellular yields for organisms which ferment the various substrates. The greatest predicted yield is for the carbohydrate-fermenting organisms, even though the carbohydrate content of the influent is only intermediate between proteins and fats. This is a reflection of the greater energy content of carbohydrates on an electron equivalent basis. The hydrogen-fermenting organisms are relatively low in concentration since little hydrogen is formed from the fermenta-

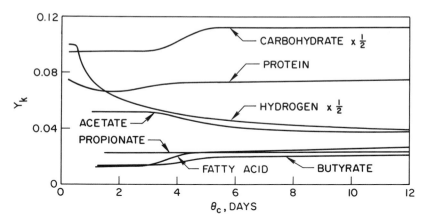

Figure 4. Computed yield coefficient as a function of cell retention time for microorganisms utilizing various substrates

tion as postulated. Although a major portion of the substrate passes through fatty acids of one type or another during the fermentation, the yield of fatty acid fermenting organisms is relatively low, again reflecting the less favorable energy release from these compounds per electron equivalent of substrate fermented.

A comparison of data from a laboratory evaluation by O'Rourke (5) of methane fermentation at 35°C and using sludge from a primary municipal waste treatment plant is illustrated in Figure 6. The digesters were operated on a semi-continuous basis, receiving waste at least once per day and more frequently than this at the lower θ_c of 2.5 days. The results are quite comparable with those obtained from the continuous feed

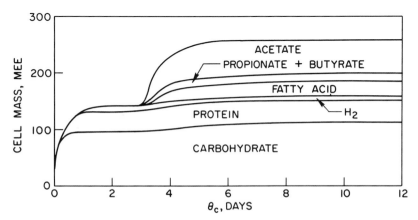

Figure 5. Computed mass of cells formed from utilization of various substrates as a function of cell retention time

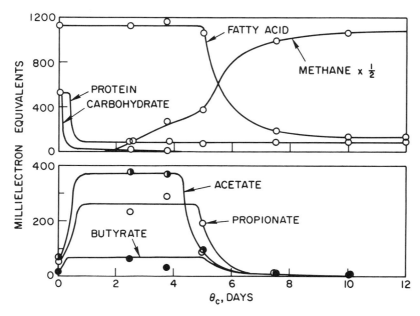

Figure 6. Experimental data from semicontinuous reactor operated at 35°C and treating primary sludge from a municipal waste treatment plant [Data from O'Rourke (5)]

model. Carbohydrate and protein degradation could not be evaluated accurately since at the lowest θ_c studied, fermentation was essentially complete. In addition, hydrogen production was not monitored so data on this constituent was not available. Butyrate production was not as high at the shorter θ_c values as predicted by the model. This suggests that the v_{ik} values assumed for butyrate formation from carbohydrates and protein in the model were high. Over-all, however, the data tend to substantiate, at least qualitatively, the results from the model analysis.

Literature Cited

(1) Bryant, M. P. *et al.*, ADVAN. CHEM. SER. (1971) **105**, 23.
(2) McCarty, P. L., "Energetics and Bacterial Growth," *Proc. Rudolf Res. Conf., 5th,* Rutgers, The State University, July 2, 1969.
(3) Hungate, R. E., "The Rumen and Its Microbes," Academic, New York, 1966.
(4) Jeris, J. S., McCarty, P. L., "The Biochemistry of Methane Fermentation Using C^{14} Tracers," *J. Water Pollut. Contr. Fed.* (1965) **37**, 178.
(5) O'Rourke, J. T., "Kinetics of Anaerobic Treatment at Reduced Temperatures," Ph.D. Thesis, Stanford University, 1968.

RECEIVED November 16, 1970.

7

Anaerobic Phase Separation by Dialysis Technique

JACK A. BORCHARDT

Department of Civil Engineering, University of Michigan,
Ann Arbor, Mich. 48104

The practical use of sludge digestion, an extremely involved biological degradation process, needs more study for adequate controls. It generally consists of two steps (phases) of biological activity—a form of β-oxidation leading to organic acid production, followed by acid decomposition leading to methane and carbon dioxide generation. Better insight into the over-all mechanism might be obtained by separating the acid-production phase from the methane-generation step through dialysis. The objective would be to feed the fastidious methane flora through a membrane in either pure or mixed culture. Such a process permits using a normal raw sludge substrate and provides an environment protected from shock. This paper describes typical equipment and experiments and indicates the type of response produced. The potential of the technique seems excellent.

The process of sludge digestion has come under a great deal of attack recently, especially in the Great Lakes region. The discharge of supernatant back to the influent of the primary tanks does produce an additional surge of mineral fertilizers through the secondary phase of treatment. As an example, one conventional activated sludge plant in Michigan was found to have peaks of ammonium salts in its effluent as high as 45 mg/liter after 6 hours of aeration. When such an effluent is discharged to a receiving water where eutrophication is a problem, the digestion phase of treatment may logically be attacked as unacceptable. Thus, more and more waste plants are using thickening and filtration of raw sludge with subsequent incineration of the cake.

This author however is of the opinion that there are conditions where digestion is particularly appropriate and that its use should not be unilaterally curtailed because of occasional improper application. Furthermore, plant modifications can be made to compensate for nutrient problems.

The conclusion which should emerge in the light of the situation outlined is that the process of digestion should not be used as an unrelated and separate operation in waste treatment. The inputs to and products of digestion should only be considered in design as part of a dynamic system in which each of the many elements comprising the whole is considered as having a cause and effect relationship in the total system response.

Sludge digestion has been casually characterized as consisting of two major phases. Phase I, generally referred to as the acid-production phase, and Phase II referred to as the gas-production phase.

Phase I is thought of as operating to decompose biochemically most raw sludge components to simple two- and three-carbon organic acids, while Phase II biologically converts these simple organic acids to methane and CO_2. The acid-producing organisms are characterized by being less sensitive and less fastidious than the gas producers. Hence when a digester is shocked by overload, toxic conditions, or extreme environmental change, the normal response of the system is to accumulate acid as the gas formers fail. Under any of these conditions, the process is marked by decreasing pH, increasing CO_2 percentages in the gas produced, accumulation of volatile acids, declining alkalinity, and declining total gas production.

This response is interpreted by engineers to mean that acid-forming organisms can tolerate adversity better than gas formers and that an accumulation of Phase I waste products (the acids) can be tolerated by acid formers to a rather marked degree.

In general, a digester is operated well below its maximum capacity in deference to the sensitivity of the gas formers. Normally too, such operation demands that design factors affecting digester volume include substantial safety factors to avoid any possibility of shock to the biological gas formers constituting Phase II of the process. Naturally the above situation constitutes a source of increased cost and loss of efficiency. If the gas formers could be protected by other means than merely providing tank volume, the process might be much more economical and the subsequent processing of supernatant liquids easier to accomplish.

The Sanitary Engineering Laboratory at the University of Michigan has for many years been investigating the digestion process by separating the phases of the process through the use of a dialyzing membrane. This paper is an attempt to relate the experiences to date to provide a background for potential users.

The problems which have been studied have been many and varied and the results not always startling. However the technique of dialysis has been very worthwhile and holds great promise for revealing some of the intricacies of the digestion process.

Membranes

One of the problems which existed when this study began was that of suitable membranes. Regenerated cellulose membranes [Visking Co., Chicago, Ill.] having an effective pore diameter of 24 A were easily available in both sheet and tube form. The problem with such material is the rate of decomposition. It was at first assumed that a run could be made before the membrane was decomposed. However, this objective proved to be impossible to attain. Usually in about 6 to 10 days the membrane would fail.

It is not inconceivable to envision two reservoirs with a separate dialysis frame. The latter could be designed to permit a new membrane to be inserted every three days or so. If the pore size of this type of membrane becomes essential for the research contemplated, the appropriate membrane holder could be built easily. Certainly there has been a vast amount of research done using regenerated cellulose. Therefore appropriate permeability constants and other necessary bits of information are much easier to acquire than would be the case for less frequently used membranes.

Much of the research in the University of Michigan laboratories has been done with the Hi-Sep 70, a vinyl plastic membrane [Graver Water Conditioning Co., New York, N. Y.] which has an effective pore size of 70 A units and is not subject to bacterial degradation. Unfortunately this material is somewhat brittle and tends to crack at points where flexure takes place in a dialysis frame. It also tends to change pore size when sterilized in an autoclave.

Quite recently our efforts have been most successful using a vinyl impregnated nylon mesh [Gelman Instrument Co., Ann Arbor, Mich.] which is now available in sheets up to 12 ft in length as well as in quilted tubes. It is available in many pore sizes and appears to stand up well under sterilization.

The technology in this area is changing rapidly as industry begins to find commercial applications, and the types of available membranes as well as the cost should become more favorable with time.

Penetration of Membranes by Microorganisms

There has been much discussion in the literature as to whether or not microorganisms can penetrate a membrane of pore structure smaller than the organism. Experience in filtration through porous media emphasizes the fact that head differential and time are essential features of the penetration phenomenon. Whether the organisms successfully deform and whether the pore size distribution plays a part by including a finite number of pores of a size larger than the mean, there does seem to be considerable evidence that membrane penetration is a distinct possibility (Figure 1).

Schultz and Gerhardt (1), after reviewing many aspects of the problem, reach the tentative conclusion that "several kinds of bacteria in fact have the ability to penetrate the pores of dialysis tubing under certain experimental conditions and that these results are not attributable to leakage." They do suggest that this finding requires confirmation by other workers.

Dialysis Culture Systems

Two types of dialysis culture systems have been used for the most part in the studies reported herein though a third type is presently under

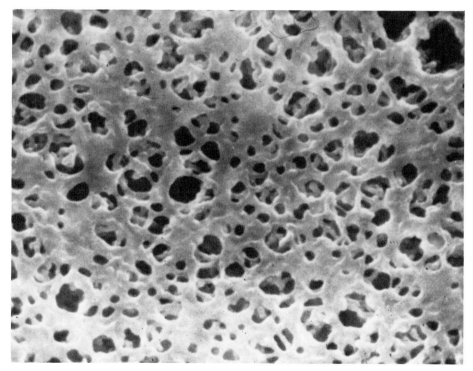

—Courtesy Gelman Instrument Co.

Figure 1. Electron microphotograph of membrane surface showing irregularity in pore structure

Figure 2.   Initial dialysis apparatus used in anaerobic work (after Hammer)

consideration. The first was of fairly large capacity and was used for
the initial study of optimum pH and electrode potentials for each of the
phases. Its design is discussed completely elsewhere (2, 3), but essentially it is based on the principle of enrichment cultures developed
through feeding techniques. This application essentially has been an
extension of the work of Gerhardt and Gallup (4, 5). These microbiologists developed a dialysis fermentor for producing large quantities of
pure cultures of microorganisms under aerobic conditions. In their application the fermentation compartment was fed through the membrane
from a nutrient source in an adjacent reservoir. Figure 2 shows our
adaptation of their concept as developed for studying anaerobic digestion.
Figure 3 gives a somewhat better perspective of the apparatus.

In concept, the operation of the dialysis unit depended on the natural problem of generating a methane flora. Because of this requirement,
the two compartments were started initially as a single unit without the
membrane in place and operated until a stable response was obtained,
indicating good methane production. The units reported on in this paper
were all fed twice a day at 9:00 am and 9:00 pm, much as the local waste
plant operator would pump sludge twice a day to his digestion equipment. When gas production in the equipment was judged to be normal,
the hoses to the dialysis frame were clamped and the membrane inserted.

Thereafter, only one side of the system was fed raw sludge. The organisms of Phase II of the process had to receive their substrate through the membrane. In this application, (Digester No. 2), the unfed side of the system, rapidly became an enrichment culture of methane organisms.

The dialysis membranes have several functions in this type of experiment. For example, the pores must be small enough to prevent the passage not only of organism but also of most of the organic molecules or fragments thereof, which might continue to support the acid-forming flora that existed initially in Digester No. 2. It was also important to prevent passage of exocellular digestive enzymes either way through the membrane. The Hi-Sep membrane with pores of 0.007 μ very closely approximated these requirements. Its diffusion coefficient (D) for acetic acid is (1.0 sq cm/sec $\times$ 10^5) while the coefficient for protein is (0.05 $\times$ 10^5 sq cm/sec) (6).

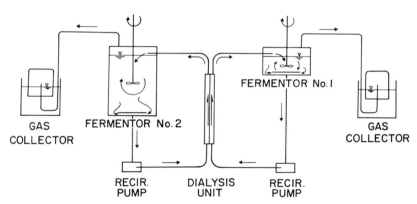

Figure 3.   Schematic of dialysis–fermentation tank apparatus

As far as other considerations were concerned, the area of membrane had to be ample to pass all the acid necessary for support of the methane formers in a fraction of the time between feedings of the digester. This requirement was necessary to permit adequate latitude in adjusting pumping cycles. A large pumping capacity in addition to the two sizes of digestion compartments in the unit permit ample flexibility for studying either acid production or gas formation. Figure 4 shows a typical curve from this research. After a series of such observations, Figure 5 can be constructed indicating definite optima (2).

It should be emphasized that in this application the methane producers in the isolation of Digester 2, were ideally supported by their natural substrate. The dialysis technique is unique in that it would flush away any waste materials as well as provide all the trace growth factors necessary for the optimum growth of this type of culture, whether or

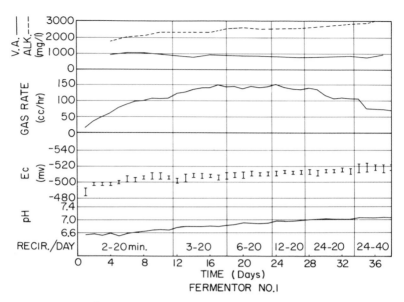

*Figure 4. Daily operational data sheet, optimum E_c and pH for acid
production at a loading of 0.16 lb vs./cu ft fermentor No 1 per day
(after Hammer)*

not these growth factors had been actually defined experimentally. It is
conceivable that the pore size of the membrane might somewhat limit the
presence of certain essential dietary ingredients, but this phenomenon
could be studied merely by changing the pore size of the membrane and
studying the culture response.

Additionally the slow growth rate of the methane formers which has
always caused some problems with wash-out or overgrowth would be
entirely eliminated without filtering, centrifuging, or in any way handling
these fastidious organisms.

In this type of experiment too, the acid-producing phase might be
expected to continue to support some methane organisms so that the
phases would not be truly separated. Several steps can be taken to help
minimize this aspect of the problem. First, the feed rate to Phase I can
be raised drastically so that wash-out is severe. Second, the sludge before feeding can be finely screened to break up all protective clumps of
organic matter, and then the mass can be aerated for a short time to
inhibit the methane flora which might be present. After taking such pre-
cautions the sludge must be allowed to become completely anaerobic
before feeding.

After performing many such experiments it has been tentatively
concluded that Phase I appears to continue to generate some methane

gas even in the absence of a demonstrable methane flora as presently conceived. We are currently studying the many ramifications of this aspect of digestion which dialysis culture makes possible.

The second type of dialysis unit being used is considerably smaller than the first type described and is designed specifically of borosilicate glass for complete autoclaving (Figure 6). The units are specially made by the glass blower, assembled with the new nylon reinforced membrane in place, and sterilized. In use, one side is charged with digesting sludge and the other with previously prepared dialysate. The latter is brought into proper equilibrium using electrode potential as the monitoring and control device whereupon the dialysate can then be seeded with pure cultures of methane organisms (supplied by Paul H. Smith, Department of Bacteriology, University of Florida).

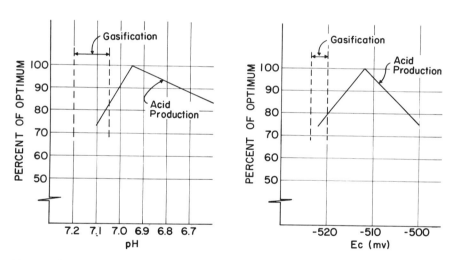

Figure 5.   Comparison between the general optimum E_c–pH conditions for acid production and the E_c–pH conditions at optimum gasification of volatile acids during digestion for raw sludge loadings in the range of 0.1 to 0.2 lb vs./cu ft/day (after Hammer)

Digester Start-Up with Dialysis Control

Whether or not digester start-up by control of the acid phase through dialysis can ever be consummated on a practical scale, the laboratory demonstration is exciting and informative. Figure 7 shows the response of a digester which was started without seed of any kind. The tank was merely filled with raw sludge (7). To make the situation even more impossible, the unit was fed raw sludge twice a day in the normal fashion. Such a digester in practice would be hopelessly stuck, and no amount of neutralizing and reseeding would be of any help (8). In this case, however, the acids generated were dialyzed away, and the unit came into proper operation in a period of two weeks. An experience of this kind poses the question: is control merely by extraction of the acid, or is some additional stimulant involved?

Dialysis against Distilled Water

The dialysis of an actively digesting sludge mass against a tank of distilled water presents a few problems. For example, though the levels of fluids in each reservoir may be kept the same, there is still an adverse pressure generated because of the difference in osmotic pressures. This produces an actual flow of water through the membrane. It can be controlled however by throttling the discharge hose on one side of the dialysis frame and building up a back pressure sufficient to offset the adverse osmotic pressure. Initially such pressure differentials are quite large; then as the distilled water takes on the character of the digesting sludge, the pressure differential disappears.

It was during the preparation of this type of dialysate that membrane penetration by microorganisms became the subject of debate. We wanted

Figure 6.   Details of borosilicate glass dialysis unit for separating phases of digestion

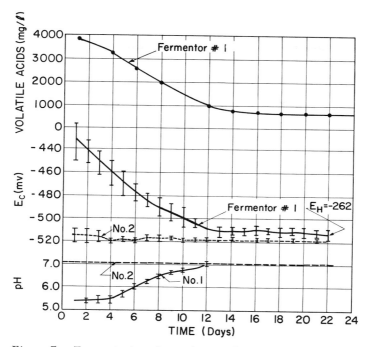

Figure 7.   Fermentation of raw sluge with raw sludge feed using optimum dialysis transfer of volatile acids

a quantity of such dialysate for study, and yet our large unit was plastic and could not be autoclaved. Time after time the attempt was made to autoclave all tubing and assemble the unit aseptically using ethylene oxide to disinfect the plastic portions. Still, after a few days of operation, in each case, the dialysate would become cloudy and was found to contain bacteria.

The final procedure adopted was to discharge the dialysate as it came past the membrane through a glass column wound with an electric heating tape. The column, fluid flow, and tape wattage were adjusted to bring the fluid to pasteurization temperature. The fluid from that time on was free of organisms. This dialysate is now held refrigerated in two-liter florence flasks, ready for inoculation with pure cultures of methane organisms in the small borosilicate glass units described above.

Electrode Potential Control in Dialysis

Hammer in his doctoral work (3) released purified hydrogen gas directly over a platinum electrode while the latter was suspended in equilibrated dialysate. Poising was non-existent, and a very nominal amount of hydrogen forced the E_c value to -641 mv, with a pH of 6.98. The electrode behaved as a hydrogen electrode, and results were predictable on a theoretical basis.

On the opposite side of the membrane the same amount of hydrogen gas produced no change in the electrode reading from its original value of E_c equal to -512 mv. Fifty times as much hydrogen gas as originally used produced a shift in E_c from -512 to -530 mv. This change could be predicted from the shift in bicarbonate equilibrium arising from the elution of CO_2 by the hydrogen.

The digesting sludge therefore has a poising capacity which is not matched in the dialysate. When sludge is dialyzed against distilled water, the alkalinity, pH, volatile acids, etc. in the dialysate equalize in about a 10 day period. During that interval, the distilled water initially tends to have

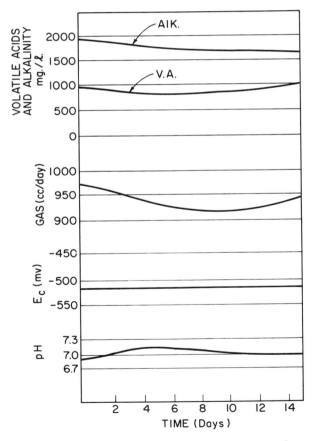

Figure 8.   Trend of variables in Digester No. 2 when Digester No. 1 is flushed with distilled water

a slightly adverse effect on the digestion. Since some of the acids are dialyzed, the pH tends to rise while the volatile acids, alkalinity, and gas rate tend to fall slightly (*see* Figure 8). About the sixth day after dialysis is begun, these components begin to rise again to a more normal level. The gas rate may tend to exceed normal for a period of time since the gas formers can be fed through the membrane as well as from the digester substrate. All this time the E_c value remains poised at a fairly steady level of -512 mv.

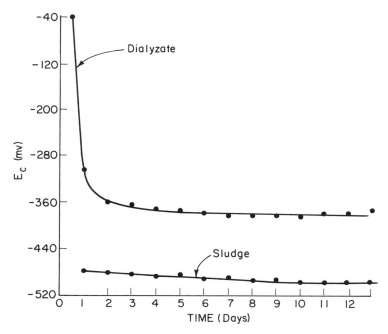

Figure 9. E_c *values on each side of membrane when dializing against distilled water*

On the opposite side of the membrane, however, the E_c value never falls below -380 mv (*see* Figure 9). This result is not caused by a lack of organic solids. When normal sludge in any stage of digestion is added to the dialysate after autoclaving, there is no change in the E_c value. Figure 10 shows this response clearly. However if that same volume of sludge is added to the dialysate without autoclaving and containing an active acid-forming culture, there is an immediate drop in potential to $E_c = -480$ mv (9).

Further, when this same type of inoculum is added after being filtered through fiberglass pads and dense Whatman filter paper and with all methane bacteria inhibited, the E_c value again drops to -480 mv.

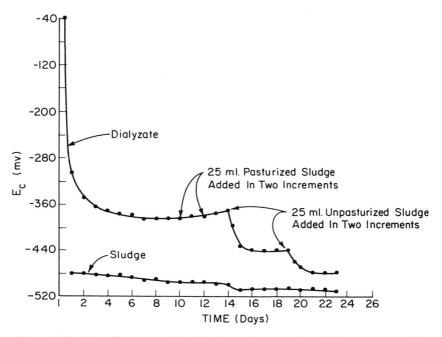

Figure 10. E$_c$ *values on each side of membrane when dializing against distilled water with pasteurized and unpasteurized sludge as inoculum*

From these results it can be concluded that the mechanism which induces
the proper E_c level representing anaerobic fermentation is non-dialysable.
It is not associated with solids which can be retained on Whatman filter
paper, it is heat labile, and has a strong poising capacity about 100 mv
below the level in equilibrated dialysate.

One further experiment has been carried out showing that the electrode response appears to be the result of active metabolism. Under
conditions where the acid formers are being fed through a membrane,
the response to stirring is sharp and precipitous. If agitation past the
membrane surface is stopped, the E_c value immediately jumps to a less
negative value (*see* Figure 11).

In the application of electrode potential measurements to anaerobic
digestion, the University of Michigan has pioneered (*2, 6, 10*). Physical
chemists have stated that the system theoretically cannot generate sufficient electroactivity to make such measurements statistically precise;
further that thermodynamic reversibility, system equilibrium, and steady-
state conditions in such biological systems are sufficiently inexact to permit potential results of meaningful nature; lastly, that sludge contaminants
are serious enough to distort materially all results (*11*).

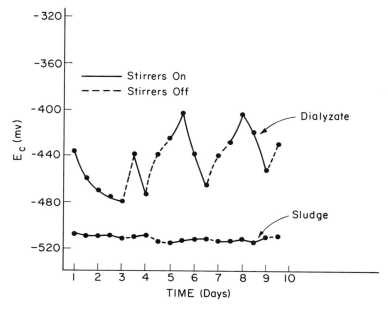

Figure 11.   *Variations in* E_c *with stirring of the dialysate*

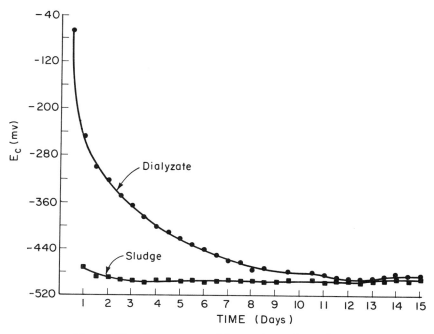

Figure 12. E_c *response with a perforated membrane*

Frankly, it is our opinion that in this area theory and practice don't agree. Electrode potential values are of great value, and much reliance must be placed on this maligned parameter. The dependence on electrode potentials in dialysis work is probably best emphasized by a typical example.

Assuming the objective is to study a pure culture of methane organisms, the first step would be to autoclave a glass dialysis apparatus similar to that in Figure 6. One side would be filled with actively digesting sludge and one side with sterile dialysate. Initial stirring is started, and the proper gas production is begun on the sludge side. Electrode readings on the dialysate side probably will read an E_c of about −380 mv. Recirculation of components past the membrane will reveal any perforations since an active growth contaminating the dialysate will drop the potential to −480 mv (*see* Figure 12). If everything remains static for a day or so, filtered sludge gas is bubbled through the dialysate until the head gases are properly adjusted. Then hydrogen is added until the E_c value is −520 mv. At that point the pure culture of methane organisms is seeded and the study begun.

No other parameter we have used can replace this tool. It reveals perforations, contaminations, bacterial activity, and presently we are convinced, even the nature of the contaminating organisms (*i.e.*, Phase I or Phase II). There is still no answer to the cause or effect status of this environmental parameter. We only know that in the digesting sludge mass it is most closely associated with active metabolism of a growing cell, and that the critical range for E_c of −480 to −520 mv would seem to be caused by an enzyme or other protein closely associated with the living cell. The study of the acid and gas phases is greatly facilitated by dialysis, and dialysis study in turn is most easily carried out using electrode potential measurements.

Having defined a range for what was believed to be an optimum value of pH and E_c for each of the two phases, it seemed logical to develop a two-stage system, poising the first at the optimum for acid formation, followed by a second stage poised at the optimum for methane generation. If optimized potential would be a sufficiently dynamic parameter to cause an accelerated growth of each type of flora and inhibition of the opposite phase would permit the development of an enrichment culture of each phase, then the need for a membrane might not exist.

The optimal values were established by an electrical poising device previously described (*10*). The system response however was strictly that of two single-stage digesters in series. No amount of acidification appeared to stop the formation of the bulk of the methane in Phase I as long as changes were made gradually. This aspect of the study seemed to indicate a rather different type of methane flora than had been previously considered characteristic of sludge digestion.

Since the first phase seemed to respond so well as a single digester at most of the pH values in the normally productive range, the experiment was enlarged, and three digesters were set up. One was used as a control, and in the other two the pH was lowered in steps using HCl fed by automatic titrimeter in one and by an electrical control in the other. The pH was monitored by a timer every 15 minutes in all digesters, and each of the controlled units was automatically adjusted if needed to the desired level.

To prevent shock and permit development of new species at each level, if such were involved, pH levels were automatically assumed at 0.2 pH unit, and after a drop the controlled units were equilibrated for one to two weeks. If the units responded successfully, the next incremental drop was programmed. Results for one year are compressed into Figures 13 and 14. The digestion units were finally operated successfully at pH 4.6 without making any change in the environment, except as noted above. There was no failure, and no stuck digester developed. In the same way these units were brought back to full gas productivity at pH 7.0.

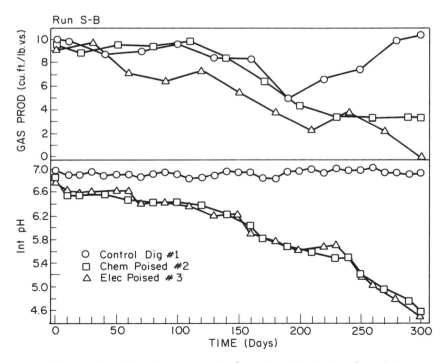

Figure 13.   Digester response to dropping pH at intervals to 4.6

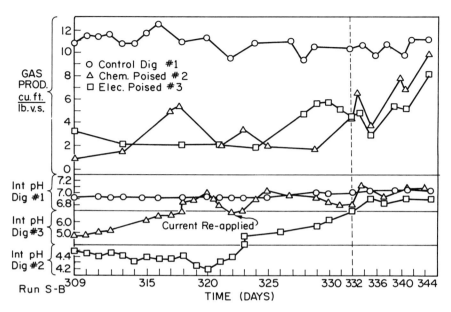

Figure 14. Digester response to raising pH at intervals from 4.6 to 7.0

At the low point the gas flow was about one-tenth maximum, the alkalinity less than 200, the volatile acids were between 1500 and 2000, and the E_c value had risen to about −470 mv in the chemically controlled unit.

Obviously though much reduced, methane production still is possible at a pH of 4.6. The reasons for stuck digesters therefore lie in other factors. Possibly too rapid a shift in environmental parameters associated with the delicate nature of the methane-producing flora should be blamed for most failure. With the amount of equilibration at each lower pH setting, it seems possible that many methane species having different optimum pH requirements may be involved in methane production and that a species which was favored at one level might not prevail at another. As long as the levels of change were gradual enough, the possibility of inducing growth of successive strains certainly is one explanation for the observations.

In conclusion, study of the organisms involved in anaerobic digestion within the protective confines of a dialysis fed reservoir can be accomplished simply, and this technique cannot help but relieve some environmental shocks to the culture. Such study may thus prevent possible erroneous conclusions relative to the organism's physiological needs and responses.

Literature Cited

(1) Schultz, J. S., Gerhardt, P., "Dialysis Culture of Microorganisms: Design, Theory, and Results," *Bacteriol. Rev.* (1969) **33**, 1–47.
(2) Hammer, M. S., Borchardt, J. A., "Dialysis Separation of Sewage Sludge Digestion," *Proc. Amer. Soc. Civil Eng.* (1969) **95** (SA5), 907–927.
(3) Hammer, M. S., "Electrode Potential-pH Relationships in the Major Stages of Sewage Sludge Digestion," Ph.D. Thesis, University of Michigan, 1964.
(4) Gerhardt, P., Gallup, D. M., "Dialysis Flask for Concentrated Cultures of Microorganisms," *J. Bacteriol.* (1963) **86**, 919–929.
(5) Gallup, D. M., "Concentrated Culture of Microorganisms in Dialysis Flask and Germentor Systems," Ph.D. Thesis, University of Michigan Libraries, 1962.
(6) Vromen, B. H., "Dialysis a Sleeper?" *Ind. Eng. Chem.* (1962) **54**, 20.
(7) Borchardt, J. A., *Proc. Intern. Conf. Water Pollut. Res., 3rd, Munich, Germany, 1966*, discussion of Paper I-13.
(8) Backmeyer, D. P., "How to Avoid Some Digesters," *Water Sewage Works* (1955) **102**, 369.
(9) Carlson, R. N., "Studies on the Electrode Potentials Developed During Anaerobic Digestion of Domestic Sewage Sludge," M.S. Thesis, University of Michigan, 1966.
(10) Dirasian, H. A., Molof, A. H., Borchardt, J. A., "Electrode Potentials Developed During Sludge Digestion," *J. Water Pollut. Control Fed.* (1963) **35**, 424.
(11) Stumm, W., "Redox Potential as an Environmental Parameter: Conceptual Significance and Operational Limitation," *Proc. Intern. Conf. Water Pollut. Res., 3rd, Munich, Germany, 1966*.

RECEIVED August 20, 1970.

8

Dynamic Modeling and Simulation of the Anaerobic Digestion Process

JOHN F. ANDREWS and STEPHEN P. GRAEF

Department of Environmental Systems Engineering, Clemson University, Clemson, S. C. 29631

One of the major problems associated with the anaerobic digestion process is its poor record with respect to process stability. Dynamic modeling and simulation are useful tools for investigating process stability and can be used to quantify operation and improve design. Some key features to be included in a dynamic model are: (1) an inhibition function to relate volatile acids concentration and specific growth rate for the methane bacteria; (2) consideration of the unionized acid as the growth limiting substrate and inhibiting agent; (3) consideration of the interactions which occur in and between the liquid, gas, and biological phases of the digester. Simulation studies can provide qualitative evidence for the validity of the model by predicting the dynamic response of those variables most commonly used for process operation.

The anaerobic digestion process is used widely for treating municipal waste sludge and is finding increasing use in treating industrial wastes which contain high concentrations of organic materials. The process has significant advantages over other methods of waste treatment. Among these are a low production of waste sludge, low power requirements for operation, and formation of a useful product, methane gas, which is a form of energy that is easily transported and stored. The digested sludge is a good soil conditioner and has some fertilizer value. It should see increasing use in land reclamation. Unfortunately, even with all these advantages the process has not enjoyed a good reputation in general because of its poor record with respect to process stability as indicated through the years by the many reports of "sour" or failing digesters. The major problems appear to lie in the area of process operation as evidenced by its more successful performance in larger cities where skilled operation is more prevalent. At present operating practice consists only of a set of empirical rules, and there is a great need for a dynamic model to put process operation on a more quantitative basis. This would lead ultimately to the development of better control procedures for preventing process failure and for optimizing process performance. A dynamic model would also be valuable in improving process design since it would allow comparison of the different versions of the process with respect to process stability. The incorporation of improved control systems in the design would also improve process stability and decrease the need for oversizing.

Most mathematical models currently used to describe the process are steady-state models and therefore cannot be used to predict process performance during start-up operations or under transient conditions resulting from changes in process inputs. Dynamic models can be used to make these predictions, and the importance of such models is realized when one considers that a failing digester is definitely not at steady state. Andrews (*1*) has developed a dynamic model for the process which is based on continuous culture theory but incorporates several important modifications. The two key features of his model are: (1) use of an inhibition function in lieu of the Monod function to relate volatile acids concentration and specific growth rate for the methane bacteria; (2) consideration of the unionized volatile acids as both the growth-limiting substrate and inhibiting agent. The use of an inhibition function is an important modification since it enables the model to predict process failure by high concentrations of volatile acids at residence times exceeding the wash-out residence time. Consideration of the unionized acids as the inhibiting agent resolves the conflict which has existed in the literature as to whether inhibition is caused by high volatile acids concentration or low pH. Since the concentration of unionized acids is a function of both total volatile acids concentration and pH, both are important. Andrews (*1*) presented experimental evidence and evidence from the microbiological literature to support his model. He also presented the results of computer simulations which provide additional supporting evidence by predicting results which have been commonly observed in field studies on digesters.

Since the process is complex and poorly understood, this first model was a simplified one with several limitations. The primary limitation was its restriction to digesters with a constant pH. The model presented in this paper removes this limitation by considering the interactions which occur in and between the liquid, gas, and biological phases of the digester. Consideration of these interactions permits the development of a model which predicts the dynamic response of the five variables most commonly used for process operation: (1) volatile acids concentration, (2) alkalinity, (3) pH, (4) gas flow rate, (5) gas composition. The results of simulations of batch cultures and the simulated responses of continuous flow cultures to step changes in substrate concentration and flow rate are presented and compared qualitatively with results commonly observed for field digesters. Even though the model still has several limitations, it should be useful in guiding experimentation and investigating the effect of different control actions and design procedures on the dynamics of the process.

Since this paper is primarily based on computer simulation, a relatively new technique, a brief discussion of this is in order. Simulation is not new since physical models have long been used to simulate processes. Many laboratory experiments are attempts to simulate physically the real system. Computer simulation has many of the same advantages and disadvantages of physical simulation. Among the advantages are considerable insight into the process mechanisms and savings in time and money. Much knowledge can be obtained about the process through the development of a mathematical model and the subsequent computer simulations using this model. Considerable monetary savings are usually realized since experimentation on the computer is less expensive than construction of a full scale plant or physical model with subsequent experimentation. Time can be compressed on the computer with simulations being conducted in minutes. This is especially important for processes such as the anaerobic digestion process where rates are slow and physical experimentation may require weeks and even months.

There are, of course, disadvantages to computer simulation. The simulation results are no better than the mathematical model on which they are based, and many first simulations therefore give only qualitative answers. Simulation with physical models does not have this disadvantage since no mathematical model is required. However, physical simulations do have the problem of scale-up from the model to the real system. This disadvantage of computer simulation can be overcome by working back and forth between mathematical modeling, computer simulation, and field observations since these complement one another. Experimentation with physical models may also be incorporated into this iterative procedure. Knowledge gained in simulation is useful for modifying the mathematical model and guiding physical experimentation. The model should be considered evolutionary in nature since it will change as more knowledge is gained about the system through simulation, field observation, or physical experimentation.

In the past, even simple models of biological processes have presented a computational bottleneck since they are largely made up of sets of nonlinear differential or partial differential equations for which analytical solutions are not usually available. However this bottleneck has been largely removed by the availability of analog computers and continuous systems modeling programs for digital computers. The simulations presented in this paper were performed using CSMP–360 (2) on the IBM 360/40 computer.

Theory

The anaerobic digestion of complex organic wastes is normally considered to consist of reactions occurring in series as illustrated in Figure 1. This is an approximation since many different microbial species may participate in the process, and their interactions are not always clearly defined. For example, recent evidence (3) indicates that some non-methanogenic bacteria may be involved in the conversion of volatile acids to methane and carbon dioxide. However, since most species of methane bacteria have much lower growth rates (4, 5) than the acid-producing bacteria, the over-all conversion of the intermediate products, volatile acids, to methane and carbon dioxide is usually considered to be the rate-limiting step in the process and will be the only reaction considered herein. The methane bacteria also appear to be more sensitive than the acid-producing bacteria to changes in environmental conditions such as pH, temperature, and inhibitory substances.

Stoichiometry

Buswell and co-workers (6) have studied the anaerobic decomposition of many organic materials and have presented a general formula (Equation 1) for the conversion of complex organic materials

$$C_nH_aO_b + \left[n - \frac{a}{4} - \frac{b}{2}\right] H_2O \rightarrow$$

$$\left[\frac{n}{2} - \frac{a}{8} + \frac{b}{4}\right] CO_2 + \left[\frac{n}{2} + \frac{a}{8} - \frac{b}{4}\right] CH_4 \quad (1)$$

to carbon dioxide and methane. However, this formula does not include the fraction of substrate that is converted to microorganisms which, although small, is necessary for the development of a dynamic model of the process. A more general formula (Equation 2) for converting organic material to microorganisms, carbon dioxide, and methane is shown below:

$$\text{Organics} \rightarrow Y_{X/S} [C_6H_{12}O_6] + \frac{1}{Y_{CO_2/X}} [CO_2]_T + \frac{1}{Y_{CH_4/X}} [CH_4] \quad (2)$$

where:

$Y_{X/S}$ = moles organisms produced/mole substrate consumed

$Y_{CO_2/X}$ = moles CO_2 produced/mole organisms produced

$Y_{CH_4/X}$ = moles CH_4 produced/mole organisms produced

Equation 3 is a more specific formula for the metabolism of acetic acid and is developed from an oxidation–reduction balance using the appropriate yield "constants" as defined.

$$CH_3COOH \rightarrow 0.02 \ C_6H_{12}O_6 + 0.94 \ [CO_2]_T + 0.94 \ CH_4 \qquad (3)$$

The empirical composition of the microorganisms is assumed to be $C_6H_{12}O_6$. The value of $Y_{X/S}$ for acetic acid was estimated from the data of Lawrence and McCarty (5). All quantities are expressed in moles for ease of manipulation in the model.

The simulations presented are for the conversion of acetic acid to microorganisms, methane, and carbon dioxide. Yields will be different for other volatile acids, especially noticeable being the increased ratio of methane to carbon dioxide produced as the length of the carbon chain increases. For field digesters utilizing complex substrates it would also be necessary to include the carbon dioxide generated by the acid-producing bacteria.

The carbon dioxide produced is indicated as $(CO_2)_T$ since it can remain dissolved in the liquid, react chemically to form bicarbonate or carbonate, be transported into the gas phase, or precipitated as carbonate. The extent to which these reactions occur will be influenced by the interactions between volatile acids, alkalinity, pH, gas flow rate, and gas composition and are considered in the model with the exception of precipitation as carbonate. The methane produced does not undergo any chemical reaction and being slightly soluble is transported quantitatively into the gas phase.

Kinetics of the Steady State

Current design practice is based mostly on empirical procedures determined from experience. However, in recent years there has been a trend toward a more rational basis for design through the application of continuous culture theory. This theory can be expressed as a mathematical model which is based on material balances of substrate and organisms for a continuous flow, complete-mixing reactor using the three basic relationships given in the following equations. Andrews (4), among others, has discussed the assumptions involved in the development of this model.

Growth Rate

$$\frac{dX}{dt} = \mu X \tag{4}$$

Monod Function

$$\mu = \hat{\mu} \left[\frac{S}{K_s + S} \right] \tag{5}$$

Yield

$$\frac{dX}{dt} = -Y_{X/S} \frac{dS}{dt} \tag{6}$$

Organism Balance

Accumulation = Input − Output + Reaction

$$\frac{dX_1}{dt} = \frac{F}{V} X_o - \frac{F}{V} X_1 + \mu X_1 \tag{7}$$

Substrate Balance

Accumulation = Input − Output + Reaction

$$\frac{dS_1}{dt} = \frac{F}{V} S_o - \frac{F}{V} S_1 - \frac{\mu}{Y_{X/S}} X_1 \tag{8}$$

where:

X = organism concentration, moles/liter
S = substrate concentration, moles/liter
F = liquid flow rate to reactor, liters/day
V = reactor liquid volume, liters
μ = specific growth rate, day^{-1}
$\hat{\mu}$ = maximum specific growth rate, day^{-1}
K_s = saturation constant, moles/liter
o = subscript denoting reactor influent
1 = subscript denoting reactor effluent

Solutions for the steady state are obtained from Equations 7 and 8 by setting the derivatives equal to zero and solving the resulting algebraic equations for substrate and organism concentrations in the reactor effluent. The effects of other factors such as endogenous respiration, utilization of substrate for maintenance energy, and organism death may also be incorporated into the model. Most models used to describe biological waste treatment processes incorporate an organism decay term in the organism balance equation to account for the disappearance of organism mass from endogenous respiration and death with lysis. The model presented in this paper is based on continuous culture theory as illustrated in Equations 4 through 8 except that an inhibition function (described later) is substituted for the Monod function (Equation 5). An organism decay term is not included, although it can be added easily, in order to keep the model as simple as possible.

Although there is a lack of data for the application of the steady-state model to design, it does give considerable insight into anaerobic digestion and other biological processes used for waste treatment. For example, the model predicts that effluent substrate concentration is independent of influent substrate concentration thus indicating that the influent substrate concentration should be as high as possible for most efficient utilization of the reactor volume. This prediction is verified by the field studies of Torpey (7) in which he was able to triple the organic loading on his digesters by thickening the raw sludge prior to its introduction into the digester. As is frequently the case, Torpey's work preceded the prediction by theory, and sludge thickening is now standard practice in many waste water treatment plants. The model also predicts that separation and concentration of the organisms with recycle of the concentrate to the reactor influent would permit substantial increases in the hydraulic loading. Although this is not yet used widely in anaerobic digestion practice, the validity of this conclusion is substantiated by the work of Steffan (8) on meat packing wastes, in which he was able to operate a field digester at hydraulic residence times of 0.5 to 1.0 days instead of the usual 10 to 20 days. The recycle of a concentrated organism suspension is also fundamental to the operation of the activated sludge process.

Current Operating Practice

Current operating practice consists of a set of empirical rules concerning such items as organic and hydraulic loading rates, uniformity of loading, temperature control, and mixing. Empirical procedures have also been developed for the start-up of digesters and recovery of digesters which have failed or are on the verge of failure. One of the greatest problems is the early detection of impending process failure so that proper control measures may be applied to prevent failure. The restarting of a failed digester requires considerable time and effort. Since the process is complex, poorly understood, and there are strong internal interactions, there is no single variable which will always indicate the onset of digester failure, and several variables must be monitored for good control. The most important of these are listed below.

Volatile Acids Concentration. The rate of increase in volatile acids concentration, a dynamic variable, has been considered as the best individual indicator of the need for control action. Since the process is a series reaction, as shown in Figure 1, this increased rate can be caused

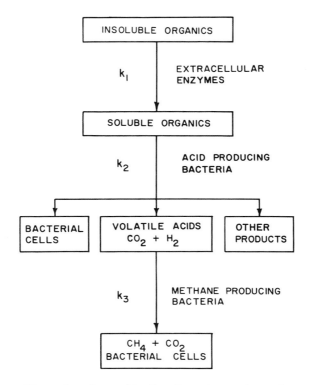

Figure 1.   Anaerobic digestion of organic wastes

by an increased rate of production of volatile acids by the acid-producing bacteria or a decreased rate of consumption by the methane bacteria. The two most common volatile acids found in digesting sludge are acetic and propionic acids, and there is some evidence that propionic acid may be more inhibitory to the methane bacteria than acetic acid. This indicates that the type of acid present could also be an important variable. Volatile acid concentrations found in well-operated sewage sludge digesters vary from 1 to 5 mmoles/liter. However, instances have been reported where digesters have been satisfactorily operated at volatile acid concentrations much higher than 5 mmoles/liter.

Alkalinity. The alkalinity in a digester serves as a buffer thus preventing rapid changes in pH. Digester stability is therefore enhanced by a high alkalinity concentration. A decrease in alkalinity below the normal operating level has been used as an indicator of pending failure since it usually precedes a rapid change in pH. Under normal operating conditions, the major buffering system is the carbon dioxide–bicarbonate system with the ammonium ion as the major cation. However, when the concentration of volatile acids is high, a substantial portion of the alkalinity as measured is caused by the volatile acids since the analysis for alkalinity is conducted by titrating with a strong acid to a pH of 4.3. The pK of the volatile acids is about 4.5; therefore, a considerable portion of the titrant is used to neutralize the volatile acids salts. The authors believe that the bicarbonate alkalinity would be a better indicator of digester condition than the alkalinity as currently measured since the bicarbonate alkalinity would have a greater rate of change than the measured alkalinity and is a more fundamental variable. Although the bicarbonate alkalinity can be calculated from the volatile acids concentration and measured alkalinity, a separate analysis, as proposed by Banta and Pomeroy (9) would be preferred. The solids in the reactor and other compounds such as phosphates and silicates can contribute to the buffering capacity of the reactor although no studies have been made to determine the significance of these substances with respect to buffering capacity.

The alkalinity depends on the composition and concentration of the feed sludge among other things. High protein wastes will have higher alkalinities because of the production of ammonia from proteins. High carbohydrate wastes will have lower alkalinities. The thickening of sludge has resulted in higher alkalinities because of the increased feed rate of protein in the sludge. Typical values for alkalinity concentrations found in domestic sludge digesters are 40 to 80 meq/liter.

pH. Another indicator of possible digester problems is a decrease in pH below its normal operating level. The pH is influenced by the other operational variables and is an important control variable since the first step usually taken when pending digester failure is detected is to control the pH near neutrality by adding a base. The start-up period can also be reduced by controlling the pH near neutrality. Unfortunately, pH is not a sensitive indicator since it is a log function and, for a digester with adequate buffering capacity, it does not decrease significantly until the digester is seriously affected. The normal pH range for the digestion of domestic sewage sludge is 6.8 to 7.4.

Gas Production Rate. A decreasing gas production rate sometimes indicates that the process is not operating properly; however, this can also be caused by non-uniform feeding or changes in the carbon dioxide content of the gas. The properties of the gas phase show more variation with respect to time than those of the liquid phase since the liquid phase, because of its larger volume and longer residence time, has more damping or buffering capacity than the gas phase. The gas produced from domestic sewage sludge should average about 1 liter/gram of volatile solids destroyed.

Gas Composition. An increase in the carbon dioxide content of the gas produced has also been used to indicate the onset of digester failure. However, there is considerable variation in this with respect to time not only because of the lower damping capacity of the gas phase but also because the carbon dioxide content can be influenced by both chemical and biological reactions in the liquid phase. The normal carbon dioxide content for gas produced from domestic sewage sludge is 30–35%.

Methane Production Rate. To the authors' knowledge this has not been used as an indicator variable although one instance (*10*) has been reported where it was proposed that the percentage of methane in the gas would be a valuable indicator of digester condition. The methane production rate should be one of the better indicators of impending operational problems since the methane bacteria represent the rate-limiting step in the process. Also, methane does not participate in any chemical reactions in the digester as does carbon dioxide and therefore would be a better indicator of the biological activity of the methane bacteria than either the total gas production rate or carbon dioxide content. The smaller volume, and therefore smaller damping capacity of the gas phase, would also be an advantage since changes in methane bacteria activity would be reflected more rapidly in the gas phase than in the liquid phase. However, this lower damping capacity could be a disadvantage in the case of non-uniform feeding.

Interactions between Variables. Obviously there are strong interactions among the variables mentioned above. As an example, consider

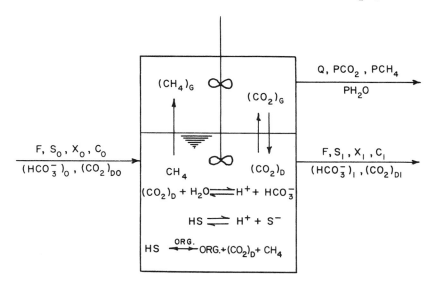

Figure 2.   Inputs, outputs, and reactions for anaerobic digester model

a digester which has received an overload of organic materials. The concentration of volatile acids will increase since the acid-producing bacteria can convert organics to volatile acids faster than they can be utilized by the methane bacteria. These acids will react with bicarbonate alkalinity to produce carbon dioxide, thus changing the gas production rate and gas composition. The increasing carbon dioxide and decreasing alkalinity will cause the pH to decrease. Both the additional volatile acids and decreased pH will result in an increase in the unionized volatile acids concentration which may rise to a high enough level to cause inhibition of the methane bacteria. Inhibition of the methane bacteria means that they cannot consume volatile acids as rapidly as before; hence, the methane production rate will decrease, and the concentration of volatile acids may continue to increase until failure occurs. This illustrates the importance of these interactions and shows the need to put these qualitative statements in the more quantitative form of a dynamic mathematical model.

Dynamic Model

The dynamic model presented herein builds on that reported previously (1) by incorporating the interactions between volatile acids, pH, alkalinity, gas production rate, and gas composition. The model is developed from material balances on the biological, liquid, and gas phases of a continuous-flow, complete mixing reactor. Appropriate relationships such as yield constants, an inhibition function, Henry's law, charge balances, and ionization equilibria are used to express the interactions between variables. The inputs and outputs for the reactor and the reactions considered are illustrated in Figure 2.

The model has been kept as simple as possible by considering the conversion of volatile acids to methane and carbon dioxide as the rate-limiting step and including only this biological reaction in the model. It is also assumed that there are no lag period, endogenous respiration, organism death, adaptation, mutation, or inhibition by substances other than volatile acids. The input to the reactor consists of a single volatile acid, acetic, instead of complex organics, and the cations are assumed to be in the reactor influent instead of being generated internally. The carbon dioxide–bicarbonate system is assumed to be the only buffering system present, and the model is restricted to a pH range of 6–8. Transport of carbon dioxide to or from the solid phase as carbonate is not considered. By using the appropriate equilibrium expressions it would be possible to extend the pH range beyond 6–8 and also consider transport to and from the solid phase. However, beyond this pH range other factors may overshadow the effects of growth rate limitation by unionized volatile acids. At the price of increased complexity the model could also be extended to incorporate many other factors, and incorporation of some of these will undoubtedly be necessary as experimental results are obtained.

Inhibition Function. Incorporation of an inhibition function into the model in lieu of the Monod function is essential for process failure to be caused by high concentrations of volatile acids at residence times exceeding the wash-out residence time as is observed in the field. Koga and Humphry (*11*) have shown that continuous cultures obeying the Monod function are stable except at the wash-out residence time. The function proposed by Haldane (*12*) for the inhibition of enzymes by high substrate concentration will be used as an inhibition function and may be expressed as:

$$\mu = \frac{\hat{\mu}}{1 + \dfrac{K_s}{S} + \dfrac{S}{K_i}} \tag{9}$$

where:

K_i = inhibition constant, moles/liter

Andrews (*13*) has discussed the basis for the use of this function, and a more detailed treatment of its properties is given by Dixon and Webb (*14*). Yano and Koga (*15*) and Edwards (*16*) have discussed the properties of other functions which may be used to reflect inhibition by substrate. Figure 3 illustrates the properties of the inhibition function with the Monod function shown for comparison. For low values of substrate concentration or high values of K_i (less inhibitory substrates) the inhibition function reduces to the Monod function. There is a considerable reduction in the maximum growth rate attainable as compared with the case without inhibition. The maximum rate attainable may be obtained by setting the first derivative of Equation 9 equal to zero and can be expressed as:

$$\hat{\mu}_m = \frac{\hat{\mu}}{1 + 2\left[\dfrac{K_s}{K_i}\right]^{0.5}} \tag{10}$$

where:

$\hat{\mu}_m$ = maximum growth rate attainable in presence of inhibition, day^{-1}.

The substrate concentration at this growth rate is:

$$S_m = (K_s K_i)^{0.5} \tag{11}$$

where:

S_m = substrate concentration at maximum specific growth rate attainable in the presence of inhibition, moles/liter.

Modifications of this function, or the use of an entirely different function, may be necessary as more knowledge is gained about the process. One of the major defects of the function is that it does not express the fact that as substrate concentrations increases, inhibition shades into toxicity which results in organism death. This would make the process more unstable.

Inhibition by Unionized Acid. Since continuous culture theory is based on the establishment of some relationship between a limiting substrate and growth rate, it is important to identify that substance which serves as the limiting substrate. Various substances act as limiting substrates, and the form in which the substrate exists is important. The model to be used here proposes that the unionized volatile acids are the growth-limiting and -inhibiting substrate for the methane bacteria. A previous paper (1) presented experimental evidence and evidence from the microbiological literature to support this hypothesis and has shown, using computer simulations, that a model incorporating growth limitation and inhibition by unionized volatile acids predicts results similar to those observed in the field. The inhibition function can be modified to reflect this by considering the unionized acid as the limiting substrate and expressing K_s and K_i as concentrations of unionized acid (Equation 13).

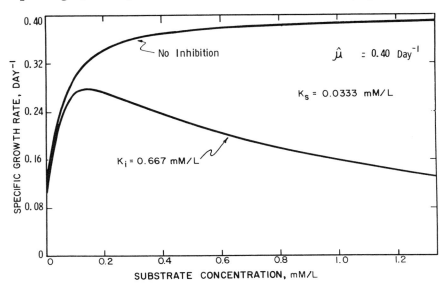

Figure 3.   Substrate inhibition function

For pH values above 6, the total substrate

$$HS \rightleftarrows H^+ + S^- \tag{12}$$

$$\mu = \frac{\hat{\mu}}{1 + \dfrac{K_s}{HS} + \dfrac{HS}{K_i}} \tag{13}$$

where:

HS = unionized substrate concentration, moles/liter.
S^- = ionized substrate concentration, moles/liter.
H^+ = hydrogen ion concentration, moles/liter

concentration, S, is approximately equal to the ionized substrate concentration, S^-, as stated in Equation 14. Therefore, at a fixed pH and with a known total substrate concentration, the unionized substrate concentration can be calculated from the equilibrium relationship for the substrate (Equation 15).

$$S = HS + S^- \text{ or } S \simeq S^- \tag{14}$$

$$HS = \frac{(H^+)(S)}{K_a} \tag{15}$$

where:

S = total substrate concentration, moles/liter.
K_a = ionization constant, $10^{-4.5}$ for acetic acid at 38°C and an ionic strength of 0.02.

The effect of pH on the inhibition function is shown in Figure 4. Low pH values greatly enhance the inhibitory effect at high substrate concentrations because of the much higher unionized substrate concentrations at these pH values. However, at lower values of total substrate concentration a low pH may have a beneficial effect because of the higher growth rates possible. Values of substrate and organism concentrations for a complete mixing, continuous-flow reactor at constant pH may be obtained by substituting Equation 13 for μ in the basic continuous culture model (Equations 7 and 8) with HS being related to S by Equation 15.

At this stage of development the model is restricted to a constant pH reactor and considers only two (pH and volatile acids concentration) of the five variables considered important for monitoring digester operation. This restriction of constant pH can be removed and the model extended to incorporate the interaction with bicarbonate alkalinity by considering the carbon dioxide–bicarbonate equilibrium as shown in Equations 16 and 17.

$$(CO_2)_D + H_2O \rightleftarrows H^+ + HCO_3^- \qquad (16)$$

$$\frac{(H^+)\,(HCO_3^-)}{(CO_2)_D} = K_1 \qquad (17)$$

where:

$(CO_2)_D$ = dissolved carbon dioxide concentration, moles/liter
(HCO_3^-) = bicarbonate concentration, moles/liter
K_1 = ionization constant, $10^{-6.0}$ at 38°C and an ionic strength of 0.2.

The hydrogen ion concentration shown in Equation 15 no longer must be considered as a constant since it can now be calculated from Equation 17 if the dissolved carbon dioxide and bicarbonate concentrations are known. It is now necessary to examine methods for calculating the dissolved carbon dioxide and bicarbonate concentrations.

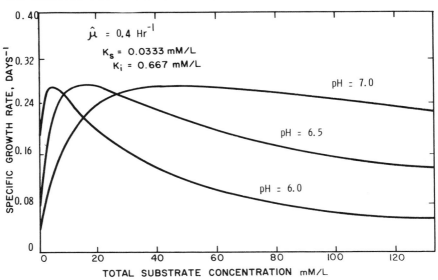

Figure 4.   *Effect of pH on substrate inhibition function*

Bicarbonate Alkalinity. The bicarbonate alkalinity in the reactor can be related to the substrate and net cation concentrations through a charge balance as given in Equation 18. For pH between 6 and 8 this charge balance can be simplified to that shown in Equations 19 and 20.

$$(H^+) + (C) = (HCO_3^-) + 2(CO_3^{2-}) + (S^-) + (OH^-) + (A) \qquad (18)$$

$$[(C) - (A)] = (HCO_3^-) + (S) \qquad (19)$$

$$(HCO_3^-) = (Z) - (S) \qquad (20)$$

where:

C = concentration of cations other than the hydrogen ion, eq/liter
A = concentration of anions other than those shown in equation 18,
   eq/liter.
Z = net cation concentration, C − A, eq/liter.

The substrate concentration, S to use in Equation 20 can be determined from the substrate balance (Equation 8). To calculate Z, it is necessary to make a material balance as shown in Equation 21.

Net Cation Balance

$$\text{Accumulation} = \text{Input} - \text{Output} + \text{Reaction}$$

$$\frac{dZ_1}{dt} = \frac{F}{V} Z_o - \frac{F}{V} Z_1 + 0 \tag{21}$$

The reaction term in this balance is zero since it has been assumed that neither C or A participate in other reactions such as precipitation or solubilization. With the bicarbonate concentration now known from Equation 20, the pH can be calculated from Equation 17 for constant values of dissolved carbon dioxide. The model is no longer restricted to a constant pH and now considers three (pH, volatile acids concentration, and bicarbonate alkalinity) of the five desired variables. However, it is restricted to a constant dissolved carbon dioxide concentration.

Gas Composition. To remove the restriction of constant dissolved carbon dioxide concentration, it is necessary to consider the interactions between the gas and liquid phases of the reactor. At equilibrium the dissolved carbon dioxide concentration can be related to the partial pressure of carbon dioxide in the gas phase by Henry's law as shown in Equation 23.

$$(CO_2)_G \rightleftarrows (CO_2)_D \tag{22}$$

$$(CO_2)_D{}^* = K \, P_{CO_2} \tag{23}$$

where:

$(CO_2)_G$ = concentration of CO_2 in the gas phase, moles/liter of gas volume
$(CO_2)_D{}^*$ = concentration of dissolved CO_2 in the liquid phase when in equilibrium with the gas phase, moles/liter
P_{CO_2} = partial pressure of CO_2 in the gas phase, mm Hg
K = Henry's law constant, $10^{-4.49}$ moles/mm Hg–liter at 38°C an ionic strength of 0.2

Calculations of the dissolved carbon dioxide concentration for use in Equation 17 from Henry's law now permits the model to consider all five desired variables except the gas flow rate. However, it is restricted to a constant partial pressure of carbon dioxide in the gas phase and equilibrium between the gas and liquid phases.

Let us now consider the strong interactions between the gas composition, bicarbonate alkalinity, and pH. These are illustrated in Figure 5 which has been developed from Equation 17 assuming equilibrium between the carbon dioxide in the gas and liquid phases. In interpreting Figure 5 remember that the bicarbonate concentration represents the difference between the net cation concentration and the volatile acids concentration as given in Equation 20. For example, a bicarbonate concentration of 50 mmoles/liter could be attained with Z equal to 50 and S equal to zero or by Z equal to 100 and S equal to 50. A high net cation concentration is therefore important in enabling the digester to resist pH changes arising from increases in the concentration of volatile acids.

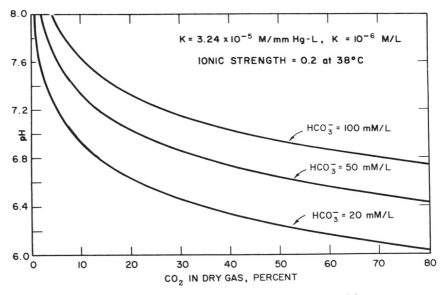

Figure 5.   Gas and liquid phase relationships at equilibrium

This may account for the difficulties frequently encountered in the anaerobic digestion of wastes containing high concentrations of carbohydrates. These digesters usually have a low net cation concentration since there is little protein present to generate the ammonium ion as in the digestion of domestic sewage sludge. Another important point illustrated by Figure 5 is that control of the carbon dioxide in the gas phase could be an effective technique in controlling the pH of the reactor. Even with a low bicarbonate concentration it would still be possible to maintain the pH near neutrality by scrubbing the carbon dioxide from the gas phase thereby reducing the percentage of carbon dioxide.

Banta and Pomeroy (9) have shown that under normal operating conditions the carbon dioxide in the gas and liquid phases can be considered to be in equilibrium. However, this may not be so under dynamic conditions since transfer across an interface is slow compared with ionic reaction rates. An expression for gas transfer, shown in Equation 24, will therefore be incorporated into the model to increase its applicability. This standard expression permits calculation of the rate of transfer of carbon dioxide either to or from the liquid phase.

$$T_G = K_L a \left[(CO_2)_D{}^* - (CO_2)_D \right] \tag{24}$$

where:

T_G = gas transfer rate, moles/liter-day.
$K_L a$ = gas transfer coefficient, day^{-1}.

Gas Flow Rate. The model can be extended to consider all five desired variables, and the restriction of a constant partial pressure of carbon dioxide can be removed by developing material balances for carbon dioxide in both the liquid and gas phases. The material balance on dissolved carbon dioxide is shown in Equation 25. R_B is the rate of production of carbon dioxide from the substrate by the methane bacteria and R_C is the rate of production of carbon dioxide from bicarbonate. The reaction of substrate and bicarbonate to produce carbon dioxide is given in Equation 28.

Dissolved Carbon Dioxide Balance

$$ \overset{\text{Biological}}{} \overset{\text{Chemical}}{} \overset{\text{Gas}}{}$$
$$\text{Accumulation} = \text{Input} - \text{Output} + \text{Reaction} + \text{Reaction} + \text{Transport}$$

$$\frac{d(CO_2)_{D1}}{dt} = \frac{F}{V} (CO_2)_{DO} - \frac{F}{V} (CO_2)_{D1} + R_B + R_C + T_G \tag{25}$$

$$R_B = Y_{CO_2/X} \, \mu X_1 \tag{26}$$

$$R_C = \frac{F}{V} (HCO_3^-)_o - \frac{F}{V} (HCO_3)_1 + \frac{dS_1}{dt} - \frac{dZ_1}{dt} \tag{27}$$

$$HS + HCO_3^- \rightleftharpoons H_2O + CO_2 + S^- \tag{28}$$

where:

R_B = biological production rate of CO_2, moles/liter–day
R_C = chemical production rate of CO_2, moles/liter–day

A material balance must also be made on the carbon dioxide in the gas phase to calculate the partial pressure of carbon dioxide for use in Henry's law. This is given (assuming complete mixing) in Equation 29.

Gas Phase Carbon Dioxide Balance

Accumulation = Input − Output + Reaction

$$\frac{d(CO_2)_{G1}}{dt} = -\frac{V}{V_G} T_G - \frac{Q}{V_G} (CO_2)_{G1} + 0 \tag{29}$$

where:

Q = total dry gas flow rate, $Q_{CH_4} + Q_{CO_2}$, liters/day
$Q_{CH_4} = DVY_{CH_4/X} \, \mu X_1$, liter/day
$Q_{CO_2} = -DVT_G$, liters/day
D = 25.5 liters gas/mole gas at 38°C
V_G = reactor gas volume, liters

The units in which the carbon dioxide in the gas phase are expressed, moles CO_2/liter of reactor gas volume, can be converted to a partial pressure of carbon dioxide by using Equation 30.

$$\frac{P_{CO_2}}{P_T} = (CO_2)_G D \tag{30}$$

The material balance on carbon dioxide in the gas phase then becomes:

$$\frac{dP_{CO_2}}{dt} = -P_T D \frac{V}{V_G} T_G - \frac{P_{CO_2}}{V_G} Q \tag{31}$$

where:

P_T = total pressure of CO_2 and CH_4 in the reactor, assumed constant at 710 mm Hg

A summary of the mathematical model showing the information flow between phases is given in Figure 6.

Simulation Studies

The studies presented herein should be considered only semiquantitative in nature since it has been necessary to make several simplifying assumptions in developing the model, and reliable values for many of the parameters are not available. Reasonable estimates of $\hat{\mu} = 0.4$ day^{-1}, $K_s = 0.0333$ mmole/liter, and $Y_{X/s} = 0.02$ mole/mole were made from the data of Lawrence and McCarty (5). For acetic acid, $Y_{CO_2/X}$ and $Y_{CH_4/X}$ are equal and were determined from the basic stoichiometry (Equation 3) as 47.0 moles/mole. An order of magnitude estimate of $K_i = 0.667$ mmole/liter was made using the 2000–3000 mg/liter of total volatile acids that Buswell (6) considers to be inhibitory. The estimates for K_s and K_i are not as reliable as those for $\hat{\mu}$ and the yield constants because K_s and K_i must be expressed as concentrations of unionized acid,

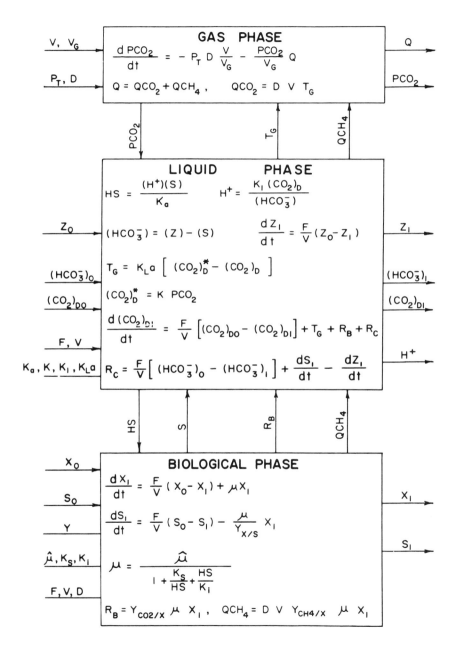

Figure 6.   Summary of mathematical model and information flow

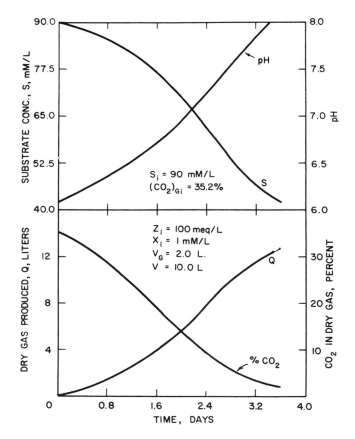

Figure 7A.   General characteristics for batch reactor

and the estimates therefore depend upon an accurate knowledge of the pH. All values are estimated for a temperature of 38°C, and the equilibrium constants are used for this temperature and an ionic strength of 0.2. Values for the gas transfer coefficient, K_La, are not available for anaerobic digesters. A value of $K_La = 100$ day^{-1} was selected to maintain the dissolved carbon dioxide concentration reasonably close to the equilibrium value since it was not the purpose of this paper to investigate the effect of gas transfer rate on the process.

Batch Reactors. The model for a batch reactor is obtained easily from the continuous flow reactor model by setting the liquid flow rate equal to zero wherever it occurs in the material balances. Simulation results showing the effects of varying the inhibition constant, initial organism concentration, and pH control will not be presented since the general effects of these at constant pH have been demonstrated previously (*1*). The results of these previous simulations indicated that in batch reactors inhibitory substrates caused pronounced lag periods. Substantial reductions in the lag period could be obtained by increasing the initial organism concentration, decreasing the initial substrate concentration, or increasing the initial pH.

Results of a simulation to illustrate the general behavior of a batch reactor are given in Figures 7A and 7B. The initial organism concentration, X_i, has been selected as 1 mmole/liter and the initial percentage of carbon dioxide as 35.2 (250 mm Hg). The net cation concentration, Z_i, and the initial substrate concentration (acetic acid), S_i, have been set at 100 meq/liter and 90 mmoles/liter, respectively. These values of Z_i and S_i are higher than would normally be used in laboratory experiments or encountered in the field but have been selected to demonstrate how the specific growth rate varies over a wide range of unionized substrate concentration. The bicarbonate alkalinity is not shown but would initially be 10 mmoles/liter and increases as substrate concentration decreases in

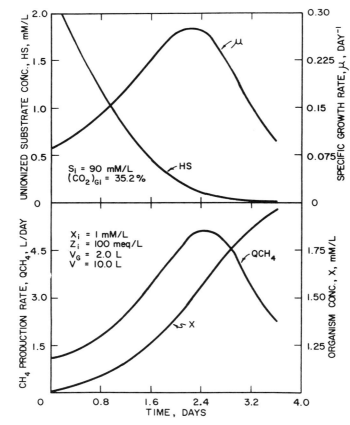

Figure 7B.   General characteristics for batch reactor

accordance with Equation 20, $HCO_3^- = Z - S$. Since Z is constant, the bicarbonate concentration would approach Z as a limit.

The substrate consumption rate is low at first not only because organism concentration is low but also because the unionized substrate concentration is high thus inhibiting the organisms and causing a low specific growth rate. As the organisms consume the substrate, the pH rises because of an increase in bicarbonate concentration and a decrease in dissolved carbon dioxide as the carbon dioxide in the gas phase is flushed out. The carbon dioxide produced by the organisms does not go into the gas phase but is converted to bicarbonate in accordance with Equation 20 which shows that for each mmole of substrate utilized a mmole of bicarbonate must be formed since the net cation concentration is constant. The decrease in substrate concentration and rise in pH result in a lower unionized substrate concentration with the specific growth rate reaching its maximum of 0.28 day^{-1} at a unionized substrate concentration of 0.15 mmole/liter as indicated in Figure 7B and as calculated from Equations 10 and 11. As the unionized substrate concentration decreases further, the specific growth rate also decreases since growth is now limited by the low unionized substrate concentration (see Figure 5). The effects of both changes in organism concentration and specific growth rate are reflected in the methane production rate, thus indicating that under some conditions this may be a better indication of digester condition than some of the more commonly used indicators.

There is a marked rise in pH as the reaction progresses thus indicating the need for careful selection of the initial conditions in experimental work on the methane bacteria. This increase is caused by the decreasing concentration of dissolved carbon dioxide as the initial carbon dioxide is flushed out of reactor and the increasing concentration of bicarbonate as substrate is consumed. This variation in pH can be decreased by starting with lower initial net cation and substrate concentrations as illustrated in Figure 8. In this instance, where Z_i and S_i are 50 and 20 mmoles/liter, respectively, the pH only increases from 6.6 to 7.2 whereas in the previous instance it increased from 6.1 to 8.0. Still another possibility for decreasing pH variation is to use a larger reactor gas volume as shown in Figure 9 where the gas volume is increased from 2.0 to 20.0 liters with the liquid volume being maintained constant at 10.0 liters. This also results in a substantial decrease in pH variation (6.6 to 6.9 as contrasted with 6.6 to 7.2) since the percentage of carbon dioxide is maintained more constant and suggests a possible means of controlling pH in field digesters. Many digester installations use gas recirculation for mixing and also have a gas storage reservoir to smooth out fluctuations in gas production rate. These reservoirs would also smooth out, or damp, fluctuations in gas composition, and recirculation of gas from the reser-

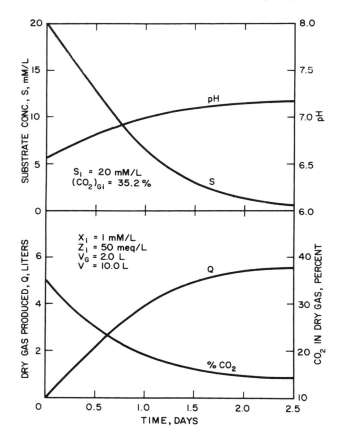

Figure 8.   Effect of initial substrate and cation concentration for batch reactor

voir to the digester could help to decrease fluctuations in digester pH thus making the process more stable. If the reservoir capacity is inadequate, it would be possible to raise the digester pH by scrubbing some of the carbon dioxide from the recirculated gas.

The model cannot be used to simulate the batch start-up of digesters without adequate seed sludge since the pH will fall below 6 under these conditions.

Continuous Flow Reactor. The general effects of an inhibitory substrate on the stability of a continuous-flow, complete-mixing reactor have also been demonstrated through simulation studies previously (*1*). The results of these simulations for controlled pH reactors showed that: (1) increasing the quantity of seed sludge or increasing the pH would decrease the time required for start-up; (2) digester failure could occur during start-up if insufficient seed sludge were present or the pH were

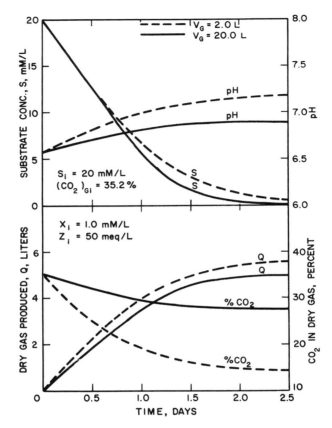

Figure 9.   Effect of reactor gas volume for batch reactor

too low; (3) digester failure could be caused by sudden changes in the organic or hydraulic loading on a digester; (4) digester failure by organic overloading during start-up could be avoided by slowly (ramp forcing instead of step forcing) bringing the digester loading to its full value; (5) pH control or decrease of process loading were effective techniques for curing failing digesters.

Since the stability of a digester is a function of the conditions prevailing in the digester at the time of load change, it is of interest to examine values of the five important operational variables at steady state. One of the most important input variables affecting these operational variables is the input net cation concentration, Z_o, and a plot of the values of these operational variables at steady state *vs.* Z_o is given in Figure 10 for a reactor with a residence time of 10 days and an input substrate concentration (acetic acid), S_o, of 167 mmoles/liter. The steady-state operating values of the substrate and bicarbonate concentrations increase as Z_o increases. These increases have been observed by Albertson (*17*), among others, in physical experiments using domestic sewage sludge. The increase in alkalinity level within a digester tended to increase the operating level of the volatile acids, and he reported volatile acids concentrations of 1.7 to 2.5 mmoles/liter at an alkalinity of about 25 meq/liter and concentrations of 8 to 10 meq/liter at alkalinities of 100 meq/liter. These are the same order of magnitude as shown in Figure 10. This figure therefore helps to explain why some digesters have been known to operate successfully at high volatile acid concentrations. High volatile acid concentrations can be tolerated if the pH is high enough to maintain the concentration of unionized acids below the inhibitory level. However, operation at a high volatile acid concentration is not desirable since this increases the load on other processes in the treatment plant which treat the effluent from the digester. In some instances it might be desirable to control the digester pH at a lower value to decrease the effluent concentration of volatile acids.

Figure 10 also shows that there is a marked decrease in the percentage of carbon dioxide in the gas phase as Z_o increases. This is because the higher input cation concentration allows more of the biologically produced carbon dioxide to be converted into bicarbonate (Equation 20) and indicates why the percentage of carbon dioxide in the gas phase is not always a good indicator of the biological activity of the digester. The total gas flow rate also decreases as more of the carbon

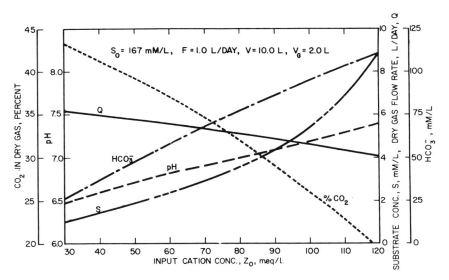

Figure 10.   Effect of influent cation concentration on steady state substrate concentration for continuous flow reactor

dioxide produced by the organisms is converted to bicarbonate. Although not shown, the decrease in methane production rate would be much less, decreasing from 3.5 liters/day at $Z_o = 30$ meq/liter to 3.3 liters/day at $Z_o = 120$ meq/liter and would therefore be a better indicator of digester conditions.

In a field digester these steady-state values would also be affected by the fact that most of the volatile acids and cations (NH_4^+) would be generated internally with some carbon dioxide being produced by the acid-producing bacteria. Acids other than acetic are also formed, and it is of interest to compare the steady state conditions for other volatile acids with those of acetic. The formulas for converting propionic and butyric acids to microorganisms, methane, and carbon dioxide are given in Equations 32 and 33.

$$CH_3CH_2COOH \rightarrow 0.025\ C_6H_{12}O_6 + 1.18\ (CO_2)_T + 1.67\ CH_4 \quad (32)$$

$$CH_3CH_2CH_2COOH \rightarrow 0.03\ C_6H_{12}O_6 + 1.41\ (CO_2)_T + 2.41\ CH_4 \quad (33)$$

The values for the organism yield constants are assumed to be somewhat higher than for acetic acid since they are expressed on a molar basis and more than one species may participate in the reaction. The yield constants for carbon dioxide and methane are developed from the oxidation–reduction balances.

Values of the important operational variables at steady state are given in Table I for acetic, propionic, and butyric acids. For comparison, the simulations were conducted with the biological parameters, μ, K_s, and K_i being the same as those previously assumed for acetic acid. The major differences between the three simulations is that as the length of the carbon chain increases, there is a decrease in the percentage of carbon dioxide and an increase in the gas flow rate. This indicates that in field digesters the gas composition and gas flow rate could be useful in indicating changes in the composition of the influent substrate. However, these changes could also be caused by changes in the net cation concentration as illustrated in Figure 10.

The validity of the model can be tested by subjecting it to simulation conditions which cause failure in field digesters, seeing if the model also predicts failure, and comparing the response of the operational variables given by the simulation with those observed in the field. Simulations of both organic and hydraulic overloading have therefore been made. In both instances the reactor was initially at steady state (*see* Table I) for an input substrate concentration (acetic acid), S_o, of 167 mmoles/liter, influent net cation concentration, Z_o, of 50 meq/liter, and a residence time of 10 days. The liquid volume, V, was 10 liters, and the gas volume, V_G, was 2.0 liters.

Process failure by organic overloading was simulated by a sudden or step change in S_o from 167 to 667 mmoles/liter, and the results are shown in Figures 11A and 11B. The volatile acids concentration and percentage of carbon dioxide in the gas phase both increase and the bicarbonate alkalinity and pH both decrease in a manner similar to that

Table I.   Steady-State Conditions for Different Volatile Acids[a]

	Acetic	Propionic	Butyric
pH	6.72	6.78	6.81
S, mmoles/liter	1.87	2.15	2.30
HCO_3^-, mmoles/liter	48.13	47.85	47.70
Q (dry), liters/day	6.51	10.60	14.65
CO_2 (dry), %	39.2	33.7	31.2
X, mmoles/liter	3.30	4.11	4.93
$Y_{X/S}$	0.020	0.025	0.030
$Y_{CO_2/X}$	47.0	47.0	47.0
$Y_{CH_4/X}$	47.0	47.0	80.0

[a] S_o = 167 mmoles/liter, C_o = 50 meq/liter, V = 10.0 liters, V_G = 2.0 liters

commonly observed in the field. However, the total gas flow rate at first increases from 6.5 liters/day to a maximum of 26.6 liters/day before starting to decrease. This increase is caused by the chemical production of carbon dioxide from bicarbonate by reaction with the volatile acids and also an increased methane production rate that results from an increased concentration of unionized acid which raises the specific growth

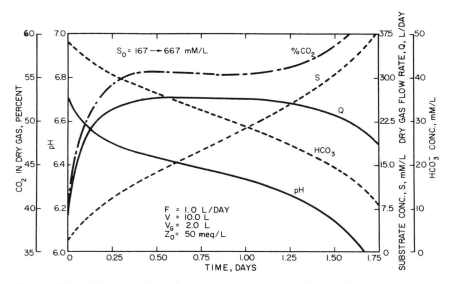

Figure 11A.   Process failure by a step change in influent substrate concentration for a continuous flow reactor

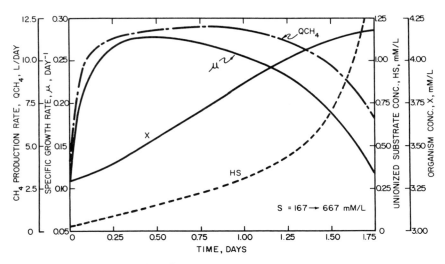

Figure 11B. Process failure by a step change in influent substrate concentration for a continuous flow reactor

rate of the microorganisms (Figure 11B). The specific growth rate reaches its maximum of 0.28/day at about 0.5 day. However, the methane production rate continues to increase after this time since the decreasing specific growth rate is offset by a continued increase in organism concentration and does not start to decrease until about 0.75 day has elapsed.

An additional simulation showed that the process would not fail for this step change in influent acid concentration if the influent net cation concentration was simultaneously forced upward in the same ratio—*i.e.*, from 50 to 200 meq/liter. The failure of a domestic sewage sludge digester from organic overloading with sewage sludge would fall somewhere between these two extremes since additional cations (NH_4^+) would be generated internally from protein and the volatile acids are also generated internally. Rates of change of the operational variables during failure would probably be somewhat less than those indicated by the simulation. The simulation would approximate failure more closely by an overload of carbohydrates since there would be no internal generation of cations and carbohydrates are rapidly converted to volatile acids.

Failure of the process by hydraulic overloading is simulated by a step change in the flow rate from 1.0 to 3.0 liters/day which changes the residence time from 10.0 to 3.33 days. The simulation results are presented in Figures 12A and 12B. This corresponds to the classic continuous culture condition of "organism washout" where the organisms are being washed out of the reactor faster than they can reproduce. Once again, all the classic symptoms of failure are shown except that the total gas production rate increases before decreasing. Comparison of Figures 11B and 12B shows that there is a distinct difference between the two types of failure. For failure caused by hydraulic overloading the organism concentration does not increase to a maximum before decreasing but immediately starts to decrease.

Failure can also be caused by the addition to the digester of inhibitory or toxic substances other than volatile acids. The model presented is not adequate to represent this type of failure. However, this could be accomplished by adding a death rate term to the organism balance (Equation 7) and establishing a material balance on the toxic material.

Although the results of these simulations are adequate to show the qualitative validity of the model, it would be premature to make quantitative conclusions until more accurate values for the parameters are established and the model is modified to reflect the effect of inhibitory or toxic substances other than volatile acids. However, some generalizations are possible. Although failure can be prevented, if detected early enough, through such control actions as pH adjustment and heavy metal precipitation, it is important to determine the cause of failure so that future occurrences may be prevented or anticipated. The simulations indicate that a possibility may exist for assignment of the cause of failure through differences in the type of response. As previously mentioned, there is a distinct difference in the response of the organism concentration to hydraulic or organic overloading. Another point of general interest is that

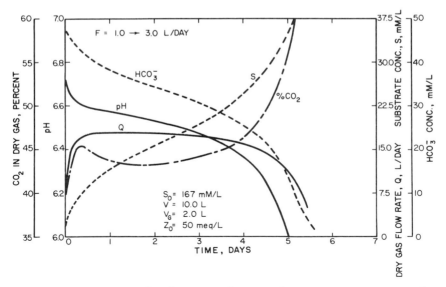

Figure 12A.   Process failure by a step change in flow rate for a continuous flow reactor

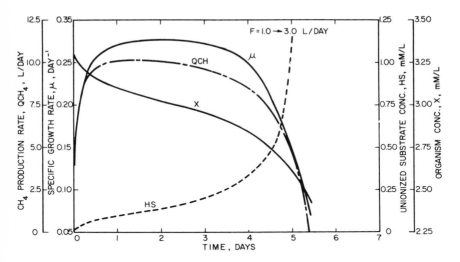

Figure 12B.   Process failure by a step change in flow rate for a continuous flow reactor

the specific growth rate curves go through a maximum in a relatively short period of time. The specific growth rate might therefore be a useful variable for indicating pending digester problems. For insoluble substrates, the specific growth rate cannot be determined since current measurement techniques cannot separate the organisms from the substrate. However, the unionized substrate concentration corresponding to the maximum specific growth rate attainable in the presence of inhibition can be calculated from pH and total volatile acids measurements. Use of the unionized acid concentration as an indicator of pending digester problems would have the advantage of incorporating the influence of all of the operational variables. As shown later, the digester will not always fail when the unionized acid concentration corresponding to the maximum specific growth rate is reached; however, it would indicate that control action may be needed.

Process stability may be enhanced by operating the reactor at longer residence times or by increasing the organism concentration in the reactor. Organism concentration can be increased by using a thickener to increase the influent substrate concentration or by separating the organisms from the effluent stream and recycling a concentrate of these organisms to the reactor. Stability may also be enhanced by operating the reactor with a higher net cation concentration. This is demonstrated in Figures 13A and 13B where again the influent substrate concentration was step forced from 167 to 667 mmoles/liter, but the reactor was initially at a steady state corresponding to an influent net cation concentration of 100 meq/

liter instead of 50 meq/liter. In this instance the reactor does not fail because of the higher buffering capacity of the system. However, the specific growth rate does reach its maximum possible value (Equation 10) and remains near this level for some time before beginning to decrease. The unionized acid concentration reaches a maximum which exceeds that (0.18 *vs.* 0.15 mmoles/liter calculated from Equation 11 before starting to decrease. In this case an unionized acid concentration exceeding that calculated from Equation 11 did not indicate pending failure.

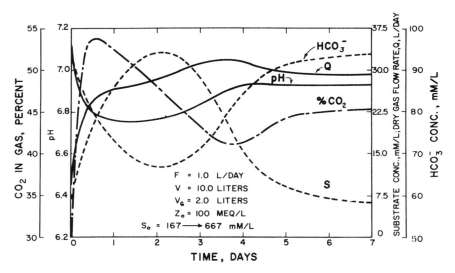

Figure 13A.   Effect of Z_o on response to a step change in influent substrate concentration for a continuous flow reactor

Additional evidence for the qualitative validity of the model can be obtained by simulating application of some of the techniques which have been used to cure failing digesters. A simplified model presented previously (*1*) predicted that a failing digester could be cured by reducing the process loading or increasing the pH in the reactor. The simulation of two other techniques for curing failing digesters, using the model presented here, are given in Figure 14. The failing digester used for these examples was initially at a steady state corresponding to an input substrate concentration, S_o, of 167 mmoles/liter and an input net cation concentration, Z_o, of 50 meq/liter. The digester was failed by a step change in S_o from 167 to 667 mmoles/liter (Figures 11A and 11B). The recovery techniques were applied at the same time the step change in loading was made. The first technique was to increase the net cation concentration in the feed from 50 to 200 meq/liter. This permitted recovery as

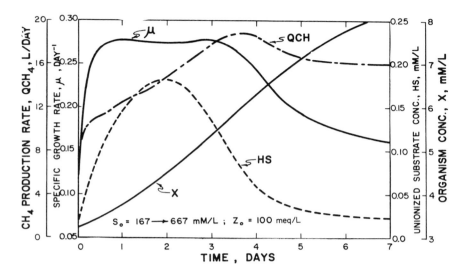

Figure 13B. Effect of Z_o on response to a step change in influent acetate concentration for a continuous flow reactor

shown. The maximum unionized substrate concentration reached was 0.18 mmole/liter at a pH of 6.60. The second technique applied was the recycle of sludge from a second stage reactor at a rate equal to 100% of the input flow rate and with an organism concentration equal to twice that in the first reactor. The recycle stream was assumed to have the same composition, other than organism concentration, as the steady-state effluent from the first stage. This technique was also effective in permitting recovery, and the maximum unionized substrate concentration reached was 0.21 mmole/liter at a pH of 6.43. The ability to recycle digested sludge for control purposes is a strong argument for the use of two-stage digestion and has long been advocated by Buswell (*18*). Although the recycled flow does increase the hydraulic loading rate to the first stage, it also dilutes the incoming waste, thereby decreasing substrate concentration, and adds additional organisms. Design of the second stage digester to concentrate the organisms more effectively before recycle would further enhance the use of recycle as a control technique.

Summary and Conclusions

The anaerobic digestion process has a poor record with respect to process stability with the major problems appearing to be in the area of process operation. There is a need for a dynamic model to quantify process operation and replace the empirical rules currently used. This model would also be of value in improving process design through comparison of the different versions of the process with respect to process stability and the incorporation of improved control systems. A dynamic model for the process is presented here. Its key features are: (1) use of an inhibition function to relate volatile acids concentration and specific growth rate for the methane bacteria; (2) consideration of the unionized acid as the growth-limiting substrate and inhibiting agent; (3) consideration of the interactions which occur in and between the liquid, gas, and biological phases of the digester. Consideration of these interactions permits the development of a model which predicts the dynamic response of the five variables most commonly used for process operation—volatile acids concentration, alkalinity, pH, gas flow rate, and gas composition.

Simulation studies provide qualitative evidence for the validity of the model by predicting results which are observed in the field. Some of these results are: (1) at steady state, an increase in the alkalinity concentration in the digester results in an increase in the operational levels of pH and volatile acids; (2) failure of the process can occur through organic or hydraulic overloading; (3) the course of failure, as evidenced by the behavior of the operational variables pH, alkalinity, volatile acids concentration, and gas composition, is qualitatively the same as that observed in the field; (4) addition of base or recycle of sludge from a second stage reactor are effective techniques for curing failing digesters.

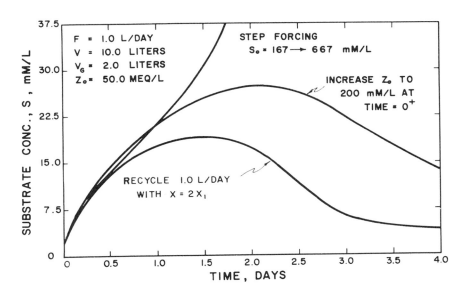

Figure 14.   Techniques for process recovery from a step change in influent acetate concentration, S_o, for a continuous flow reactor

Although the results of the simulations should not be used to draw quantitative conclusions, they do permit some generalizations. The simulations of batch reactors indicate that care should be used in selecting initial conditions for experimentation on methane bacteria to avoid marked changes in the batch fermenter during the reaction. They also indicate that it may be possible to control pH in field installations by controlling the carbon dioxide content of recirculated gas used for mixing. The simulations for continuous flow reactors indicate that it may be possible to determine the cause of failure through differences in the type of response. Another possibility is the use of the unionized acid concentration, as calculated from the total acid concentration and pH, as an improved indicator of pending digester failure. All of these points are, of course, subject to experimental verification.

Undoubtedly the model will require modification as experimental data are obtained and more comparisons are made between simulation results and field operations. However, the model should serve as a valuable framework for these modifications. Major modifications which may be needed are: (1) incorporation of a mathematical expression for the death of microorganisms by toxic materials; (2) the effect of the solid phase on reactor buffering capacity; (3) modification of the biological portion of the model to incorporate the acid-producing organisms, the effects of organism decay, adaptation, mutation, and delay of organism response to changes in substrate concentration. Work is in progress on a model incorporating some of these modifications.

The model will also be useful in guiding field experimentation and investigating the effect of different control actions and design procedures on the dynamics of the process. Decisions as to the best operational variable, or combination of variables, for control purposes can be made by simulating various control actions using deviations of these variables from their desired state as error signals. The best control action, or combination of control actions, to be taken could also be determined from these simulations. Simulations of the different versions of the digestion process would enable design engineers to compare these with respect to process stability and select that version most suitable for a given installation. Work is in progress on the simulation of different control systems and the comparison of the different versions of the process with respect to process stability.

Nomenclature

A = concentration of anions other than HCO_3^-, CO_3^{2-}, S^-, and OH^-, equivalents/liter

C = concentration of cations other than the hydrogen ion, equivalents/liter

$(CO_2)_D$ = dissolved carbon dioxide concentration, moles/liter

$(CO_2)_D{}^*$ = concentration of dissolved CO_2 in the liquid phase when in equilibrium with the gas phase, moles/liter

$(CO_2)_G$ = concentration of CO_2 in the gas phase, moles/liter of gas volume

D = conversion factor, liters gas/mole gas

F = liquid flow rate to reactor, liters/day

H^+ = hydrogen ion concentration, moles/liter

$(HCO_3{}^-)$ = bicarbonate concentration, moles/liter

HS = unionized substrate concentration, moles/liter

K = Henry's law constant, moles/mm Hg-liter

K_a = ionization constant for acetic acid, moles/liter

K_i = inhibition constant, moles/liter

$K_L a$ = gas transfer coefficient, day^{-1}

K_s = saturation constant, moles/liter

K_1 = ionization constant for bicarbonate, moles/liter

P_{CO_2} = partial pressure of CO_2 in the gas phase, mm Hg

P_T = total pressure of CO_2 and CH_4 in the reactor, mm Hg

Q = total dry gas flow rate, $Q_{CH_4} + Q_{CO_2}$, liters/day

Q_{CH_4} = $D V Y_{CH_4/X} \mu X_l$, liters/day

Q_{CO_2} = $-D V T_G$, liters/day

R_B = biological production rate of CO_2, moles/liter-day

R_C = chemical production rate of CO_2, moles/liter-day

S = substrate concentration, moles/liter

S^- = ionized substrate concentration, moles/liter

S_m = substrate concentration at maximum specific growth rate attainable in the presence of inhibition, moles/liter

t = time, days

T_G = gas transfer rate, moles/liter-day

V = reactor liquid volume, liters

V_G = reactor gas volume, liters

X = organism concentration, moles/liter

$Y_{CH_4/X}$ = yield constant, moles CH_4 produced/mole organisms produced

$Y_{CO_2/X}$ = yield constant, moles CO_2 produced/mole organisms produced

$Y_{X/S}$ = yield constant, moles organisms produced/mole substrate consumed

Z = net cation concentration, C − A, eq/liter

μ = specific growth rate, day^{-1}

μ = maximum specific growth rate, day^{-1}

μ_m = maximum growth rate attainable in the presence of inhibition, day^{-1}

1 = subscript denoting reactor effluent

o = subscript denoting reactor influent

Literature Cited

(1) Andrews, J. F., "Dynamic Model of the Anaerobic Digestion Process," *J. Sanitary Eng. Div., Proc. Am. Soc. Civil Eng.* (1969) **95**, SA1, 95.

(2) "System/360 Continuous System Modeling Program (360A-CX-16X) User's Manual." Publication H20-0367-2, International Business Machines Corp., New York, 1968.

(3) Bryant, M. P., Wolin, G. A., Wolin, J. J., Wolfe, R. S., "*Methanobacillus omelianskii*, A Symbiotic Association of Two Species of Bacteria," *Arch. Mikrobiol.* (1967) **59**, 20.
(4) Andrews, J. F., Cole, R. D., Pearson, E. A., "Kinetics and Characteristics of Multistage Methane Fermentation," Sanitary Engineering Research Lab., University of California, Berkeley, 1964.
(5) Lawrence, A. W., McCarty, P. L., "Kinetics of Methane Fermentation in Anaerobic Waste Treatment," Civil Engineering Dept., Stanford University, Palo Alto, Calif., 1967.
(6) Buswell, A. M., "Anaerobic Fermentations," *Ill. State Water Surv., Bull.* **32** (1939).
(7) Torpey, W. N., "High-Rate Digestion of Concentrated Primary and Activated Sludge," *Sewage Ind. Wastes* (1954) **26**, 479.
(8) Steffan, A. J., "Treatment of Packing House Wastes by Anaerobic Digestion," in "Biological Treatment of Sewage and Industrial Wastes," Vol. II, Reinhold, New York, 1958.
(9) Banta, A. P., Pomeroy, R., "Hydrogen-Ion Concentration and Bicarbonate Equilibrium in Digesting Sludge," *Sewage Works J.* (1934) **6**, 234.
(10) Babcock, R. H., "Instrumentation and Control in Water Supply and Wastewater Disposal," Reuben H. Donnelley Corp., New York, 1968.
(11) Koga, S., Humphrey, A. E., "Study of the Dynamic Behavior of the Chemostat System," *Biotech. Bioeng.* (1967) **9**, 375.
(12) Haldane, J. B. S., "Enzymes," Longmans, London, 1930.
(13) Andrews, J. F., "A Mathematical Model for the Continuous Culture of Microorganisms Utilizing Inhibitory Substrates," *Biotechnol. Bioeng.* (1968) **10**, 707.
(14) Dixon, M., Webb, E. C., "Enzymes," 2nd ed., Academic, New York, 1964.
(15) Yano, T., Koga, S., "Dynamic Behavior of the Chemostat Subject to Substrate Inhibition," *Biotechnol. Bioeng.* (1969) **11**, 139.
(16) Edwards, V. H., "The Influence of High Substrate Concentrations on Microbial Kinetics," 158th Meeting, ACS, New York, Sept., 1969.
(17) Albertson, O. E., "Ammonia Nitrogen and the Anaerobic Environment," *J. Water Pollution Control Federation* (1961) **33**, 978.
(18) Buswell, A. M., "Septic Tank to Controlled Digestion," in "Biological Treatment of Sewage and Industrial Wastes," Vol. II, Reinhold, New York, 1958.

RECEIVED August 3, 1970.

Application of Process Kinetics to Design of Anaerobic Processes

ALONZO W. LAWRENCE

Department of Environmental Engineering, Cornell University,
Ithaca, N. Y. 14850

A rational approach to designing biological waste treatment processes employing suspended cultures of microorganisms is developed from kinetic concepts and continuous culture theory. Steady-state models are presented for two completely mixed process configurations. Biological solids retention time (θ_c) is identified as an independent design parameter. Based on the rate-limiting step concept, methane fermentation controls anaerobic treatment of complex organic wastes. Published (1970) values of growth and substrate utilization coefficients for methane fermentation of long and short chain fatty acids, the most prevalent precursors of methane, are summarized. The suggested design approach, incorporating these kinetic coefficients, is illustrated by considering the influence of the following factors on process design: (1) kinetic failure, (2) effluent quality, (3) volumetric waste utilization rate, (4) solids recycle, (5) fermentation temperature, and (6) environmental parameters.

Anaerobic waste treatment is an important biological waste treatment process that has long been used in stabilizing municipal sewage sludges. More recently, there is increasing interest in using this process to treat high and medium strength soluble and colloidal industrial wastes. This process may be used in both packed bed reactors and the more conventional fluidized culture reactor or digester (1).

Despite widespread use of anaerobic treatment, optimum process performance is seldom achieved because a high degree of empiricism still prevails in the design and operation of such systems. A rational basis for process analysis and design is essential to the realization of the full potential of anaerobic treatment. Process kinetics—*i.e.*, the rates of waste utilization attainable and the factors affecting these rates—is an essential element in such a rational approach. Process kinetics has been used in developing: (1) information that is generally applicable to all anaerobic methanogenic fermentations of complex organic wastes (*2, 3*), and (2) information that is waste or system specific (*4, 5, 6*). The former approach, emphasized here, appears to be more productive. The success of the generalized kinetic approach depends heavily on an understanding of the microbiology and biochemistry of the anaerobic process. This is true because it is assumed that the over-all kinetics of a complex process such as anaerobic treatment is controlled by the kinetics of a process rate-limiting step.

In considering the application of process kinetics to anaerobic system design, this paper describes: (1) the kinetic model employed, (2) the microbiological and biochemical evidence used as the basis for selecting the process rate-limiting step, (3) available kinetic information for that step, and (4) an illustrative example of kinetic based process design.

Basic Kinetic Equations and Treatment Models

Basic Equations. The relationship between microbial growth and substrate utilization can be formulated in two basic equations (*2, 7*). Equation 1 describes the relationship between net rate of growth of microorganisms and the rate of utilization of the growth limiting substrate:

$$dX/dt = Y(dF/dt) - bX \qquad (1)$$

in which dX/dt = net growth rate of microorganisms per unit volume of reactor, mass/volume-time; Y = growth yield coefficient, mass/mass; dF/dt = rate of microbial substrate utilization per unit volume, mass/volume-time; b = microorganism decay coefficient, time^{-1}, and X = microbial mass concentration, mass/volume. Equation 2 relates rate of substrate utilization to both concentration of microorganisms in the reactor and concentration of the growth limiting substrate surrounding the organisms:

$$dF/dt = \frac{kSX}{K_s + S} \qquad (2)$$

in which k = maximum rate of substrate utilization per unit weight of microorganisms (occurring at high substrate concentration), time^{-1}; S = concentration of the growth limiting substrate surrounding the microorganisms, mass/volume; and K_s = half velocity coefficient, equal to the substrate concentration when $(dF/dt)/X = (1/2)k$, mass/volume. Equations 1 and 2 can be combined to yield a relationship between net

specific growth rate of microorganisms (μ) and the concentration of
growth limiting substrate, S:

$$\mu = \frac{YkS}{K_s + S} - b \tag{3}$$

in which $\mu = (dX/dt)/X$, time^{-1}. van Uden (8) has described the use
of this equation in studies of pure culture microbial systems.

Two kinetic-based process control and design parameters for the
steady-state operation of biological systems have been developed from
Equations 1, 2, and 3. These parameters are specific utilization (U)
which is equal to $(dF/dt)/X$—*i.e.*, the rate of substrate utilization per
unit mass of microorganisms; and, biological solids retention time (θ_c)
which is equal to the reciprocal of net specific growth rate (μ)—*i.e.*, the
average time period a unit of microbial mass is retained in the system.
Similarly defined parameters of empirical origin—*i.e.*, process loading
factor or food to microorganism ratio and sludge age, respectively—have
been used in wastewater treatment practice for some time. In mixed
culture microbial systems, particularly if nonbiological solids are present,
θ_c appears to be more useful for process design and control. The advantage in using θ_c rather than U is best illustrated by considering the following operational definition of θ_c:

$$\theta_c \equiv \frac{X_\tau}{(\Delta X/\Delta T)_\tau} \tag{4}$$

in which $X_\tau =$ total active microbial mass in the treatment system, mass;
and $(\Delta X/\Delta T)_\tau =$ total quantity of active microbial mass withdrawn daily,
including those solids purposely wasted as well as those lost in the effluent,
mass/time. Thus by making the reasonable assumption that the suspended
solids in the system are homogeneous, it is possible to maintain the desired steady state value of θ_c according to Equations 3 and 4 by removing
total suspended solids from the biological system at the appropriate rate
defined by Equation 4.

The efficiency of a waste treatment process is defined as:

$$E = \frac{100(S_0 - S_1)}{S_0} \tag{5}$$

in which $E =$ treatment efficiency, percent; $S_0 =$ influent waste concentration, mass/volume; and $S_1 =$ effluent waste concentration, mass/
volume. There are two efficiencies of interest: the specific efficiency,
E_s and the gross efficiency, E_g. Specific efficiency refers to the removal
of some specific component or group of components in the waste
stream. With anaerobic treatment, one may be interested in the removal

of one or more volatile acids or lipid material. Gross efficiency refers to the removal of waste as measured by some nonspecific or gross parameter —e.g., carbonaceous organic material measured as volatile solids or chemical oxygen demand (COD).

Treatment Models. Mathematical models of continuous growth pure culture microbial systems were originally presented by Monod (9) and Novick and Szilard (10). With increasing frequency, similar models have been used to describe biological waste treatment processes (7, 11, 12). A critical assumption in applying such models to complex microbial systems is that microbial growth is limited by the availability of one substance or category of substance. All other growth requirements (i.e., inorganic nutrients and trace organic growth factors) are present in excess amounts. For heterotrophic organisms, organic waste constituents are usually the growth-limiting substance while for autotrophic organisms the inorganic energy source is considered to limit growth.

Table I.   Summary of Steady-State Relationships for

Characteristic	Without Recycle

Specific efficiency

$$E_s = \frac{100(S_0 - S_1)}{S_0} \tag{5}^a$$

Effluent waste concentration

$$S_1 = \frac{K_s[1 + b(\theta_c)]}{\theta_c(Yk - b) - 1} \tag{6}$$

Microorganism concentration in reactor

$$X = \frac{Y(S_0 - S_1)}{1 + b\theta_c} \tag{7}$$

Excess microorganism production rate (mass/time)

$$P_x = \frac{YQ(S_0 - S_1)}{1 + b\theta_c} \tag{9}$$

Hydraulic retention time (V/Q)

$$\theta = \theta_c \tag{10}$$

Solid retention times

General

$$(\theta_c)^{-1} = \frac{YkS_1}{K_s + S_1} - b \tag{12}^b$$

Limiting Minimum

$$[\theta_c{}^m]_{\text{lim}} = (Yk - b)^{-1} \tag{13}$$

Maximum Volumetric Utilization

$$(\theta_c{}^*)^{-1} = Yk[1 - \left(\frac{K_s}{K_s + S_0}\right)^{1/2}] - b \tag{14}$$

---

[a] Equation numbers are given in parentheses

For a given waste treatment process configuration, models can be developed by combining the basic kinetic expressions (Equations 1 through 5) with materials balances on substrate and microbial mass for the system. While the mass balances are written in differential form, it is common to use the steady-state form of these expressions—i.e., $dX/dt = 0$ or $dS/dt = 0$. This approach appears to be quite satisfactory for design considerations.

Detailed development of steady-state models for biological treatment processes employing fluidized cultures of microorganisms have been presented by Pearson (12) and Lawrence and McCarty (11). Table I summarizes the equations which constitute such models for the two widely used anaerobic process flow configurations shown in Figure 1. The conventional sewage sludge digester is simulated by the completely mixed reactor without microbial solids recycle, while the "anaerobic activated sludge" is simulated by the reactor system with microorganism recycle.

Completely Mixed Biological Waste Treatment Processes

With Recycle

$$E_s = \frac{100(S_0 - S_1)}{S_0} \tag{5}a$$

$$S_1 = \frac{K_s[1 + b(\theta_c)]}{\theta_c(Yk - b) - 1} \tag{6}$$

$$X = \frac{Y(S_0 - S_1)}{1 + b\theta_c}\left(\frac{\theta_c}{\theta}\right) \tag{8}$$

$$P_x = \frac{YQ(S_0 - S_1)}{1 + b\theta_c} \tag{9}$$

$$\theta = \theta_c[1 + r - r(X_r/X)] \tag{11}$$

$$(\theta_c)^{-1} = \frac{YkS_1}{K_s + S_1} - b \tag{12}b$$

$$[\theta_c{}^m]_{lim} = (Yk - b)^{-1} \tag{13}$$

---

b Equation 12 is identical to Equation 3 since $(\theta_c)^{-1} = \mu$ and $S_1 = S$ for completely mixed systems at steady state.

Equation 6 shows that in both process configurations the effluent waste concentration is a univalued function of only one variable—*i.e.*, θ_c. Selection of a value of θ_c, in essence microbial growth rate, determines the concentration of growth-limiting substrate surrounding the microorganisms and in the system effluent. Figure 2 illustrates this relationship between θ_c and effluent concentration (S_1) and also treatment efficiency (E_s). An equally important relationship for process design and control is given as Equation 13. The limiting minimum biological solids retention time $[\theta_c^m]_{\text{lim}}$, is shown to be a characteristic parameter of the microbial flora in the system since it is a function of three kinetic coefficients which are themselves characterizing "constants" for a given microbial population. The value of $[\theta_c^m]_{\text{lim}}$ represents the lower kinetic limit for the steady state operation of the process. Operation of the treatment system at a lower value of θ_c will lead to removal of the microorganisms at a rate greater than their maximum growth rate. The result is microbial popu-

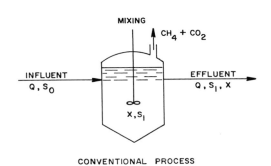

CONVENTIONAL PROCESS

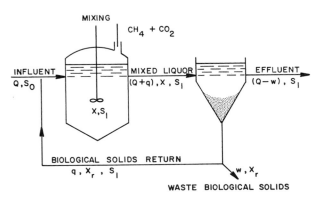

Figure 1.   Schematic of two anaerobic waste treatment processes

lation washout and system failure as defined by a waste treatment efficiency equal to zero. As shown in Figure 2, operation at values of θ_c close to $[\theta_c{}^m]_{lim}$ is incompatible with waste treatment objectives which usually specify a high treatment efficiency. The ratio of the specified operating value of θ_c to the value of $[\theta_c{}^m]_{lim}$ for a particular process can be designated as a kinetic safety factor.

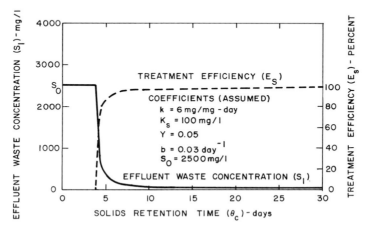

Figure 2. Steady-state relationships between solids retention time, effluent waste concentration, and specific treatment efficiency for completely mixed, continuous flow biological treatment processes

Equation 10 shows that for the conventional process (no recycle) the value of θ_c is equal to the hydraulic retention time (θ). Thus, effluent quality and treatment efficiency for a given waste are determined by the choice of a value of θ. The ability to vary θ_c independently of θ in the recycle process (Equation 11) is the principal advantage of the recycle process compared with the conventional process. This capability allows one to maintain a certain value of θ_c and hence microbial growth rate chosen according to the desired effluent quality by varying the rate of return of microorganisms to the reactor. This can be accomplished with a short θ and hence small reactor volume.

The Rate-Limiting Step Approach

Anaerobic treatment, a multistep complex process, can be described from a kinetic viewpoint as a three-step process involving (1) hydrolysis of complex organic material; (2) organic acid production; (3) methane fermentation (2). In the first step complex organics are converted to less complex soluble organic compounds by enzymatic hydrolysis. In the second step these hydrolysis products are fermented to simple organic compounds, predominantly volatile fatty acids, by a group of facultative and anaerobic bacteria collectively called "acid formers." In the third step the simple organic compounds are fermented to methane and carbon dioxide. This third step has generally been assumed to be accomplished by a group of substrate-specific, obligate anaerobic bacteria called the "methane formers." According to presently accepted concepts, the fermentation of the long chain fatty acids constitutes an important exception to this three step model. These acids are assumed to be fermented directly by methanogenic bacteria *via* a β-oxidation scheme to acetic acid and methane (*13*).

Recently, Toerien and Hattingh (*14*) have suggested that the fermentation of propionic, butyric, and long chain fatty acids to acetate and methane may not be accomplished directly by methanogenic bacteria. Rather, they hypothesize that these fermentations may be accomplished by a symbiotic association of a nonmethanogenic hydrogen releasing organism with hydrogen oxidizing methanogenic bacteria. Their hypothesis is based on the recent report by Bryant *et al.* (*15*) that *Methanobacterium omelianskii*, heretofore believed to be a pure species of methanogenic bacteria responsible for the fermentation of ethanol to acetate and methane, is a symbiotic association of a nonmethanogenic hydrogen releasing organism with hydrogen oxidizing methanogenic bacteria. The implications raised by the possible existence of such a microbial association as regards the methane fermentation kinetic studies reported to date (1970) are evaluated in a subsequent section. Here, in summary, it suffices to say that anaerobic treatment effectively converts organic waste materials to bacterial protoplasm and the gaseous end products, methane and carbon dioxide.

Given the complex stepwise nature of the anaerobic process, it is proposed that the rate-limiting step be defined as that step in the process which will cause process failure to occur under imposed conditions of kinetic stress. Kinetic stress is applied to the system by continually reducing the value of θ_c until the limiting value of θ_c—*i.e.*, $\theta_c{}^m = (\mu^m)^{-1}$, is exceeded and washout of the microbial flora results. The parameter $\theta_c{}^m$ is the value of θ_c obtained from Equation 12 when $S_1 = S_0$. The value of $\theta_c{}^m$ approaches the value of $[\theta_c{}^m]_{\lim}$ when $S_o >> K_s$.

In anaerobic treatment, failure of this type is usually evidenced by the near cessation of methane production and decreased COD removal. Several investigators (*5, 16, 17*) have reported that kinetic failure is also characterized by a build-up in the concentration of long and short chain fatty acids, the predominate precursors of methane. McCarty (*7*) and O'Rourke (*3*), in laboratory digestion studies on primary sewage sludge conducted at 35°C, confirmed the fact that the fermentation of short and long chain fatty acids to methane and carbon dioxide is the rate-limiting step in anaerobic treatment. Figure 3 shows the results obtained by O'Rourke at a fermentation temperature of 35°C, the widely accepted "optimum" temperature for mesophilic digestion. Based on his study, O'Rourke concluded that a similar situation prevails for digestion of sewage sludge at 25°C and 20°C. At 15°C, he found that lipids were not degraded significantly even at a θ_c value of 60 days. Thus it appears reasonable to conclude that over the temperature range of 35°–20°C a general solution to anaerobic treatment kinetics can be attained by delineating the kinetics of the fermentation to methane and carbon dioxide of the long and short chain fatty acids.

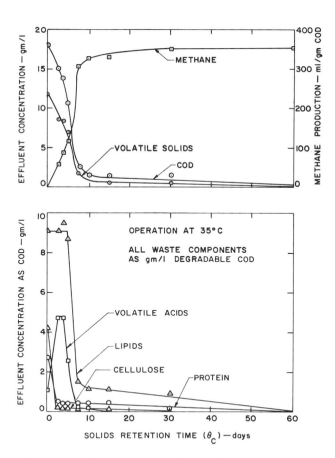

Figure 3.   Effect of θ_c on the relative breakdown of degradable waste components and methane production at 35°C (3)

Methane Fermentation Kinetics

Pursuing the rate-limiting step approach, Lawrence and McCarty (2) used laboratory enrichment cultures developed from sewage sludge to study the fermentation to methane and carbon dioxide of three important volatile fatty acid intermediates. Acetic, propionic, and butyric acids were chosen for study because (1) acetic acid has been shown to be the precursor of approximately 70% of the methane formed in the complete treatment of a complex waste; (2) acetic and propionic acids together are the precursors of 85% of the methane formed from a complex waste (13); (3) butyric acid is the precursor of an additional 8% of the methane formed from a complex waste (18). Using a similar experimental approach, O'Rourke (3) studied the fermentation to methane and carbon dioxide of long chain fatty acids and a complex waste (municipal sewage sludge) of known composition. In these studies (2, 3) the effects of temperature on the kinetic coefficients were evaluated over the ranges 35°–25°C and 35°–15°C, respectively. Fermentation temperature is of particular practical importance because large amounts of externally supplied heat could be required to raise the waste to the "optimum" temperature of 35°C. If reasonable rates of methane fermentation are obtainable at temperatures as low as 20°–25°C, anaerobic treatment could provide an attractive alternative for treating relatively low strength–high volume wastes.

Volatile Acids Kinetics. In evaluating the methane fermentation kinetics of the three volatile acids chosen for study, it was necessary to consider the process biochemistry and stoichiometry. According to Barker (19), acetic acid is fermented to methane and carbon dioxide in a single step while both propionic and butyric acids are fermented in two steps. In the first step these acids are fermented to acetic acid and methane by species of methanogenic bacteria. The resulting acetic acid is then fermented by different methanogenic species to methane and carbon dioxide. The stoichiometry of these fermentations is shown by the following equations (19).

Acetic Acid
$$CH_3COO^- + H_2O \rightarrow CH_4 + HCO_3^- \tag{15}$$

Propionic Acid
First Step
$$CH_3CH_2COO^- + \frac{1}{2} H_2O \rightarrow CH_3COO^- + \frac{3}{4} CH_4 + \frac{1}{4} CO_2 \tag{16}$$

Second Step
$$CH_3COO^- + H_2O \rightarrow CH_4 + HCO_3^-$$

Over-all

$$CH_3CH_2COO^- + \frac{3}{2} H_2O \rightarrow \frac{7}{4} CH_4 + \frac{1}{4} CO_2 + HCO_3^- \qquad (17)$$

Butyric Acid
First Step

$$CH_3CH_2CH_2COO^- + HCO_3^- \rightarrow 2CH_3COO^- + \frac{1}{2} CH_4 + \frac{1}{2} CO_2 \quad (18)$$

Second Step

$$2 CH_3 COO^- + 2H_2O \rightarrow 2CH_4 + 2HCO_3^-$$

Over-all

$$CH_3CH_2CH_2COO^- + 2H_2O \rightarrow \frac{5}{2} CH_4 + \frac{1}{2} CO_2 + HCO_3^- \qquad (19)$$

Kinetic coefficients were determined for the stabilization to methane of acetic acid and for the first step—*i.e.*, assimilation or disappearance of the feed stream volatile acid species, of the fermentation of propionic and butyric acids. The values of growth yield coefficient (Y) were computed on the basis of mg of biological solids produced per mg of substrate COD converted for energy—*i.e.*, to methane. Expressed on this basis the values of Y at a given temperature were relatively constant for the three volatile acids fermented according to Equations 15, 16, and 18. Values of the organism decay coefficient (b) were similarly constant. While some variation with temperature was noted in the values of Y and b, it was not considered significant, and at least for engineering design purposes, Y and b can be considered to be unaffected by fermentation temperature. Table II shows the range of values and the average values of Y and b for all substrates and temperatures studied ($n = 11$).

Table II. Range and Average Values of Y and b in Methane Fermentation of Volatile Acids (2)

Parameter	Range	Average
Y (mg/mg)	0.040–0.054	0.044
b (day^{-1})	0.010–0.040	0.019

Table III shows the average values of the coefficients k and K_s for the three volatile acids at the temperatures studied. Values of k are given both as the maximum specific rate of assimilation of the primary volatile acid substrates (Equations 15, 16, and 18) expressed as equivalent amounts of acetic acid and as the maximum specific rate of conversion of the primary volatile acid substrate COD to methane. The fraction of

Table III.   Average Values of Substrate Utilization

35°C

Volatile Acid Substrate	k (mg/mg–day)		K_s (mg/l)	
	"removed" as HAc^a	as COD to CH_4	as HAc	as COD
Acetic	8.1	8.7	154	165
Propionic	9.6	7.7	32	60
Butyric	15.6	8.1	5	13

[a] HAc = acetic acid

volatile acid substrate converted to methane is equal to the fraction of the substrate COD converted for microbial growth energy (20, 21). The values of K_s are expressed as equivalent concentrations of acetic acid and also as COD concentrations. The values of the coefficients in Tables II and III can be used in Equation 6 to predict the effluent concentration of each individual volatile acid. For wastes with a low lipid content, volatile acids constitute the major fraction of biodegradable organics in the effluent, and the sum of the individually computed effluent volatile acid concentrations (as COD) can be used as an estimate of the total effluent biodegradable COD for the given value of θ_c.

Table IV shows $[\theta_c^m]_{lim}$ values computed for the volatile acids using Equation 13 and $[\theta_c^m]_{lim}$ values computed or estimated for other methane fermentations discussed in subsequent sections. At 35°C the $[\theta_c^m]_{lim}$ values for acetic and propionic acids are essentially equal and differ from the values for long chain fatty acids by less than one day. Hence at 35°C the methane fermentations of long and short chain fatty acids are equally limiting in the anaerobic process. At lower temperatures, long chain fatty acid degradation appears to be the limiting phenomenon.

Table IV.   Minimum Values of θ_c for Methane Fermentation of Various Substrates

Energy Substrate	$[\theta_c^m]_{lim}$ (days)				Reference
	35°C	30°C	25°C	20°C	
Acetic acid	3.1	4.2	4.2	—	(2)
Propionic acid	3.2	—	2.8	—	(2)
Butyric acid	2.7	—	—	—	(2)
Long chain fatty acids	4.0	—	5.8	7.2	(3)
Hydrogen	0.95^a	—	—	—	(22)
Sewage sludge	4.2^b	—	7.5^b	10^b	(3)
Sewage sludge	2.6^c	—	—	—	(17)

[a] 37°C
[b] Computed values of θ_c^m for S_o = 18.1 grams/liter COD
[c] Experimentally observed value

Coefficients for Acetic, Propionic, and Butyric Acids (2)

30°C				25°C			
k (mg/mg–day)		K_s (mg/l)		k (mg/mg–day)		K_s (mg/l)	
"removed" as COD				"removed" as COD			
as HAc	to CH$_4$	as HAc	as COD	as HAc	to CH$_4$	as HAc	as COD
4.8	5.1	333	356	4.7	5.0	869	930
—	—	—	—	9.8	7.8	613	1145
—	—	—	—	—	—	—	—

Figure 4 shows the effect of fermentation temperature on K_s for acetic acid. The resulting equation, of the form of the Arrhenius equation, which describes this relationship is:

$$\log \frac{(K_s)_2}{(K_s)_1} = 6980 \left(\frac{1}{T_2} - \frac{1}{T_1} \right) \tag{20}$$

in which T_2, T_1 = temperature (°K). O'Rourke (3) extrapolated this relationship to 20°C for acetic acid and also used it as justification for developing similar relationships for the K_s values of propionic acid, long chain fatty acids, and a complex waste. While Lawrence (20) did not feel that the observed variation in values of k given in Table III was sufficient to define a temperature dependency, O'Rourke (3) developed a temperature effect relationship for Lawrence's values of k for acetic acid. He also developed temperature relationships for long chain fatty acids and a complex waste (sewage sludge). McCarty (23) has recently suggested a theoretical basis for a temperature dependency of the value of k. O'Rourke's interpretation of the temperature dependency of k and K_s for methane fermentation kinetics appears reasonable for engineering design purposes, particularly since he was able to verify experimentally the derived relationships with a reasonable degree of success.

Lipid (Long Chain Fatty Acid) Kinetics. O'Rourke (3) evaluated the methane fermentation kinetics of lipids (long chain fatty acids) in a complex waste (municipal sewage sludge). His laboratory scale experiments were operated on a semicontinuous feed basis at several values of θ_c and at temperatures of 35°, 25°, 20°, and 15°C.

He demonstrated that the lipid fraction of the raw sewage sludge accounted for approximately 50% of the biodegradable COD of the sludge. Further he showed that hydrolysis of the triglycerides, which accounted for 66% of the lipid COD, was essentially complete at values of θ_c less than the values of θ_c required for effective methane fermentation of the long chain fatty acid hydrolysis products. Stearic and palmitic

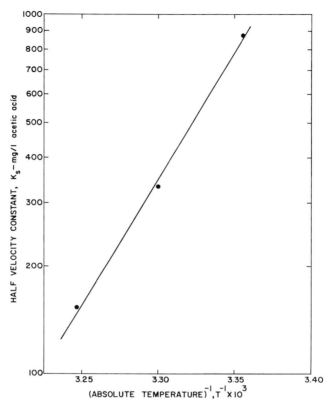

Figure 4.   Temperature dependence of K_s *in the methane fermentation of acetate* (2)

were the principal (> 66/wt %) long chain fatty acids present in the
fermentation with shorter chain saturated acids and unsaturated 18-car-
bon chain acids present in small concentrations.

Because it was not possible experimentally to determine directly
values of specific utilization (U) for fermentation of lipids (long chain
fatty acids) in digesting sewage sludge, a computational technique was
used. Values of U were calculated for each θ_c studied from a form of
Equation 3—*i.e.,* $\theta_c^{-1} = YU\text{-}b$, and assumed values of $Y = 0.04$ mg microbial solids per mg of lipid COD converted to methane and $b = 0.015$
day^{-1}. These values of Y and b are similar to the average values shown
in Table II. The use of a common value of Y for the growth of the
methanogenic bacteria responsible for fermenting long and short chain
fatty acids is an appropriate procedure since the thermodynamic free
energy "theoretically available" for microbial growth per gram substrate

Table V.   Summary of Kinetic Coefficients Developed for Lipid Material in Municipal Sewage Sludge (3)

Substrate	Temperature °C	K_s mg/l as COD	k, day^{-1}
Long chain fatty acids	20	4620	3.85
Long chain fatty acids	25	3720	4.65
Long chain fatty acids	35	2000	6.67

COD fermented to methane is the same for all the short and long chain fatty acids (20, 21). Table V shows the values of k and K_s for lipid (long chain fatty acid) fermentation.

Novak and Carlson (24) studied the anaerobic degradation of five long chain fatty acids at 37°C in enrichment cultures developed from municipal sewage sludge. Each culture was continuously fed one long chain fatty acid as the sole source of organic carbon and inorganic nutrient salts. Y, b, k, and K_s were determined for the anaerobic degradation to methane of myristic, palmitic, stearic, oleic, and linoleic acids. While these coefficients for individual fatty acids cannot be compared directly with the coefficients reported by O'Rourke (3) for lipid degradation in sewage sludge digestion, the magnitude of the values reported in the two studies appears to be significantly different. This is particularly true for k and K_s. The reasons for this lack of agreement are not apparent but appear to be related to the differences in the microbial environment, the form of the substrate, and the method of computing the coefficients. Further inquiry into the kinetics of anaerobic degradation of long chain fatty acids will be required to resolve more fully the differences in reported kinetic coefficients. O'Rourke's values are used in this paper because said values were substantiated by his experimental digestion studies on sewage sludge and consequently are more likely to be free of possible limitations imposed by an enrichment culture environment.

Hydrogen Oxidation Kinetics. Shea *et al.* (22) studied the kinetics of methane fermentation by an enrichment culture of lithotrophic (autotrophic) hydrogen oxidizing methanogenic bacteria at 37°C. Reported values of the kinetic coefficients are as follows: (1) $Y = 0.043$ mg volatile suspended solids per mg of hydrogen COD removed, (2) $b = -0.009$ day^{-1}, (3) $k = 24.8$ mg hydrogen COD removed per mg volatile suspended solids per day and (4) $K_s = 569$ mm of mercury, hydrogen pressure.

Shea's study is of particular interest in light of the postulate (14) that the fermentation to methane of propionic, butyric, and long chain fatty acids may involve hydrogen oxidizing methanogenic bacteria. If this is true, the limiting reactions in the enrichment studies by Lawrence and McCarty (2) (propionate and butyrate) and O'Rourke (3) may have been the turnover of the substrate acid by nonmethanogenic hydrogen liberating bacteria or, as appears equally likely, a self-limiting symbiotic relationship between the nonmethanogenic and the hydrogen oxidizing methanogenic bacteria. Shea's kinetic coefficients and the application of thermodynamic concepts to microbial growth as developed by McCarty (21) offered an opportunity to assess the likelihood of such a microbial symbiosis. Two aspects of such a fermentation sequence are worthy of brief note here. These deal with the predicted and observed hydrogen pressures in anaerobic fermentations and the growth yield of hydrogen oxidizing methanogenic bacteria. The fermentation of propionic acid can be used to illustrate these points.

Calculations based on the kinetics of Lawrence and McCarty (2) and Shea (22) indicate that at a 10-day value of θ_c a steady-state hydrogen pressure of 7×10^{-2} atm (53 mm Hg) would be required to assure conversion of the assimilated propionate (COD basis) to methane. Shorter values of θ_c would require correspondingly greater hydrogen pressures. Hydrogen pressures of this magnitude are seldom observed in operating digesters. Secondly, thermodynamic considerations suggest that hydrogen pressures on the order of 10^{-4}–10^{-6} atm would be required for it to be energetically possible to experience evolution of molecular hydrogen associated with the turnover of propionate under anaerobic conditions. There is also some evidence that hydrogen pressures on the order of 10^{-2} atm are toxic to the organisms mediating the turnover of propionate and will retard the fermentation of acetate to methane (18, 22). Although Shea's results on methanogenic hydrogen oxidation appear to be at variance with the postulated requirements for nonmethanogenic hydrogen evolution, it is conceivable that propionate converting bacteria could effect a biologically mediated transfer of hydrogen to the hydrogen oxidizing methanogenic bacteria at hydrogen pressures low enough to satisfy the free energy considerations mentioned above.

A second problem involved in evaluating the relevance of hydrogen oxidizing methane bacteria to anaerobic fermentation of wastes relates to the carbon source for synthesis. The value of Y reported by Shea is consistent with autotrophic growth of the organisms. However, at the hydrogen pressures suggested for propionate turnover, the autotrophic growth yields would be much smaller than indicated by Shea because of a decrease in available free energy per unit of hydrogen COD fermented to methane. This difficulty may be resolved by the report of Bryant (25) that hydrogen oxidizing methanogenic bacteria found in the rumen appear to have a synthesis requirement for carbon in the form of acetate.

Given the postulated thermodynamic requirement for low hydrogen pressures and the possible heterotrophic carbon requirement for synthesis of the hydrogen oxidizing methanogenic bacteria, the microbial explanation of the turnover of propionate could possibly be provided by the two organism theory as well as Equation 16. Such an alternate explanation would not be inconsistent with the observed kinetics (2). Similar arguments could be advanced for butyrate turnover and β-oxidation of long chain fatty acids. Resolution of these matters rests with the increasing numbers of microbiologists who are finding the microbiology and ecology of methanogenesis a fascinating and formidable undertaking.

As shown in Table IV the value of $[\theta_c^m]_{lim}$ reported for the hydrogen oxidizing culture was 0.95 day. This is considerably less than the values shown in Table IV for $[\theta_c^m]_{lim}$ for the volatile acids, lipids, and sewage sludge. Based on this information, it appears reasonable to conclude that the methanogenic bacterial oxidation of hydrogen will not be the kinetic rate limiting step *per se* in anaerobic treatment. Rather, it can be concluded that the kinetics of anaerobic treatment are governed (rate limited) by the direct methanogenic fermentation of acetic acid and the fermentations of propionic acid and the long chain fatty acids to acetic acid and methane (or hydrogen). Regardless of the actual microbial mechanism involved in these latter fermentations (propionic and long chain fatty acids), the elucidated kinetics (2, 3) appear to continue to offer a valid engineering description of the kinetics of the limiting step in converting complex organic wastes to methane and carbon dioxide.

Process Design

Biological solids retention time (θ_c) has been suggested in this paper as the kinetic based parameter of choice for use in design and control of fluidized culture continuous flow biological processes. The value of θ_c selected for design of the process, (θ_c^d), directly determines the volume of reactor needed for a conventional digester system and a given waste flow (Q) since the value of θ_c is equal to the hydraulic retention time (θ). The relationship between θ_c and the reactor volume for the system with recycle is more complex and involves consideration of the effects of solids recycle rate and recycle solids concentration.

In selecting a value of θ_c^d, the designer will be guided by consideration of two and perhaps three different θ_c values which characterize his fermentation. These are the minimum value, θ_c^m, at which process failure will occur—*i.e.*, $S_1 = S_0$ in Equation 12; the value of θ_c required to produce a specified effluent quality or treatment efficiency (Equation 12); and, in the case of the reactor without organism recycle, the θ_c value at which maximum volumetric rate of waste utilization is achieved, θ_c^*, (Equation 14). In addition to these factors the designer will have to consider the effects of fermentation temperature on kinetics and the size of the process reactor. For recycle systems, he will also have to evaluate the characteristics of the solids separation process and its effect on the over-all performance of the treatment system. Finally, non-kinetic or environmental effects such as mixing and toxic effects caused by materials added or produced in the fermentation process must be considered. These six factors which influence the selection of the design value of θ_c are discussed in more detail below.

Kinetic Failure. Based on the information in Table IV, appropriate values for $\theta_c{}^m$ at 35°C are approximately 3 days for low lipid content wastes and 4 days for high lipid content wastes (sewage sludge). At lower temperatures correspondingly larger values of θ_c are indicated. At 25°C the appropriate value for low lipid wastes is about 4 days while approximately 7.5 days are indicated for high lipid content wastes. At 20°C available information indicates a $\theta_c{}^m$ for high lipid wastes of 10 days. Essentially no lipid degradation was observed at 15°C (3). The design value $\theta_c{}^d$ will always be greater than $\theta_c{}^m$ by some "safety factor" ($\theta_c{}^d/\theta_c{}^m$) to ensure process stability against: (1) waste surges which may temporarily reduce the θ_c, and (2) environmental upsets which may temporarily retard the bacterial growth rate and, in effect, increase the microbial θ_c above the operating θ_c. O'Rourke (3) has shown that at 35°C the value of $\theta_c{}^*$ for sewage sludge in the conventional process is uncomfortably close to the value of $\theta_c{}^m$. Hence, in the interest of process stability, a somewhat longer value of $\theta_c{}^d$ would be chosen even though the dominant design objective might be to maximize the volumetric waste utilization rate. As shown in Figure 2, values of θ_c required for a high efficiency of waste treatment are considerably larger than $\theta_c{}^m$ and hence favor process stability.

Predicting Effluent Quality. Since the preponderant contributors to the biodegradable COD in a reactor effluent are the fatty acids (3, 7), the total biodegradable COD in the reactor effluent at a given θ_c value can be approximated by summing the values of S_1 for each fatty acid present, as calculated by Equation 6. For low lipid wastes, the total effluent COD would be essentially contributed by acetic and propionic acids. For wastes such as sewage sludge, the COD contribution of the long chain fatty acids which constitute approximately 35% of the biodegradable COD in this waste must also be considered. When non-biodegradable organic material occurs in the waste, the COD of such material must be added to the effluent biodegradable COD to attain the total effluent COD. O'Rourke (3) has summarized available information on the non-biodegradable fraction of the volatile solids in sewage sludge digested at 35°C. The approximate average value for this fraction was 0.45. While information on the fraction of non-biodegradable COD is not generally available, it is probably comparable with the volatile solids value.

O'Rourke (*3*) modified the above approach for computing the estimate of biodegradable effluent COD by assuming that the values of Y, b, and k for all the fatty acid fermentations could be considered equal and that Equation 6 could be modified accordingly as follows:

$$(S_1)_\tau = \frac{[1 + b(\theta_c)]}{\theta_c(Yk - b) - 1}(K_c) \tag{21}$$

in which $K_c = \Sigma K_s$ for all the fatty acids found or microbially produced in the waste in significant concentrations. Table VI summarizes the values of k; K_s for the major contributors to effluent biodegradable COD, and $K_c = \Sigma K_s$ for sewage sludge at temperature of 20°, 25°, and 35°C. Table VII shows the computed values of biodegradable effluent COD for a complex waste treated at 35°C using the values from Table VI and assumed values of Y and b of 0.05 and 0.01, respectively. Based on the total effluent COD values shown in Table VII, it is suggested that a θ_c of 10 days is the appropriate design choice for this system. Beyond 10 days, the improvement in effluent quality does not appear to be commensurate with the required increase in reactor volume. This is particularly true since effluents from anaerobic processes will usually require further treatment before ultimate disposal. Similar reasoning for treatment of a complex waste at 25° and 20°C leads to recommended values for $\theta_c{}^d$ of approximately 20 and 40 days, respectively. Lawrence and McCarty (*20*) suggested values for $\theta_c{}^d$ of 10 days at 35°C and approximately 30 days at 25°C for low lipid content wastes. Use of these recommended values in design provides safety factors in the range of 2.5–4. The large values of θ_c at lower temperature emphasize the value of the process configuration employing organism recycle.

Table VI. Summary of Kinetic Coefficients for Complex Waste (*3*)

Temp. °C	k day^{-1}	$K_s(a)$ mg/l as COD	$K_s(p)$ mg/l as COD	$K_s(L)$ mg/l as COD	K_c mg/l as COD
35	6.67	164	60	2000	2224
25	4.65	930	1140	3720	5790
20	3.85	2130	3860	4620	10610

Volumetric Utilization and Solids Recycle. VOLUMETRIC UTILIZATION. Somewhat at crossed purposes with the "effluent quality" or efficiency of treatment criteria for selecting values of $\theta_c{}^d$ is the other possible treatment objective of maximizing volumetric waste utilization (treatment) rate.

Table VII.   Prediction of Effluent Degradable COD for Treatment of Complex Waste in "Conventional" Anaerobic Digesters at 35°C

θ_c days	Effluent COD–mg/l From Indicated Sources[a]			
	Acetic Acid	Propionic Acid	Long Chain Fatty Acids	Total
4	566	211	6920	7697
6	185	69	2260	2514
8	111	42	1360	1513
10	80	30	980	1090
12	64	24	778	866
15	49	18	600	667
30	25	10	300	335
60	14	5	174	193

[a] $S_1 = \dfrac{(1 + b\theta_c)}{\theta_c(Yk - b) - 1} (\Sigma K_s)$

This objective is of special interest for concentrated wastes where per cent reduction in waste strength is more important than effluent quality since additional treatment will be required before ultimate disposal. The solids retention time at which maximum volumetric waste utilization occurs can be uniquely determined for a conventional process system (no organism recycle) according to Equation 14:

$$(\theta_c^*)^{-1} = Yk \left[1 - \left(\frac{K_s}{K_s + S_0}\right)^{1/2}\right] - b \qquad (14)$$

Since θ_c^* will occur closer to θ_c^m than the value of θ_c chosen according to the effluent quality criteria described in the previous section, the designer will usually compromise between the two ends of the design spectrum. For treating sludges by the conventional process, increased volumetric utilization at a given value of θ_c can be achieved by concentrating the waste by thickening or flotation before introduction into the digester. To aid the designer in assessing these alternatives for a conventional process system, O'Rourke (3) constructed "organic loading and decision curves" to portray graphically the tradeoffs between waste treatment efficiency, volumetric utilization, and θ_c. Figure 5 shows a curve for sewage sludge treated at 35°C (3). By way of illustration, the dotted lines on this figure indicate that a change in influent COD from 20 to 30 grams/liter at a 10-day value of θ_c will increase treatment efficiency by 3% and volumetric utilization by 70 lbs COD/10^3 ft^3/day.

SOLIDS RECYCLE. The relationships developed above do not apply to the second process scheme in which a portion of the suspended solids separated from the reactor effluent stream are returned to the reactor. Equation 11 indicates that the value of θ_c for a given waste flow is a function not only of reactor volume but also the ratio (r) of recycle flow rate (q) to raw waste flow rate (Q) and the ratio of the concentrations of recycle flow microbial solids to mixed liquor microbial solids. With this process, a unique definition of maximum volumetric utilization is not possible. The general strategy is to maintain a small value of θ and as large a value of θ_c as is compatible with treatment efficiency objectives and solids separation capability. It is possible to describe a family of curves for a given θ_c and hence effluent quality which relates recycle ratio, reactor volume, microbial solids concentration in the recycle stream, and mixed liquor solids concentration. Figure 6 shows a series of such curves for a hypothetical treatment situation. By choosing values of r and recycle flow microbial solids concentration (X_r), one can solve graphically for volume or mixed liquor solids concentration and then calculate the other quantity from Equation 8.

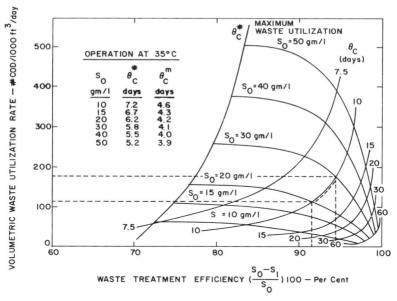

Figure 5.   Relationships between volumetric waste utilization rate, treatment efficiency, solids retention time, and raw waste degradable COD concentration at 35°C (3)

Until recently, interest in the recycle process was limited to the treatment of high strength–low solids industrial wastes such as meat packing wastes (*1*). Such systems encountered solids separation problems arising from buoyancy imparted by rising fermentation gas bubbles. Vacuum degasification has solved this problem, and hence the "anaerobic activated sludge" process is feasible for industrial waste applications.

Recently Pfeffer (26) and Torpey and Melbinger (27) have described benefits associated with recycle of digested sludge in anaerobic digestion of municipal sewage sludge. The reported increases in process performance is compatible with the prediction of the kinetic approach. Thus it appears that increasing interest will be directed toward solids recycle processes.

The ultimate in solids recycle is "total" solids retention. This is the principle of the anaerobic trickling filter recently described by McCarty (1). By providing growth adherence surfaces for the microbial population, large "effective values" of θ_c can be achieved at short hydraulic retention times. This anaerobic process application is attractive for treating relatively dilute wastes at temperatures approaching ambient conditions. Preliminary modeling of this process using some of the kinetic coefficients reported in this paper has been recently described (28). Additional study of this process is needed to establish appropriate rational design procedures.

Fermentation Temperature. The effects of fermentation temperature on performance of anaerobic biological processes have been described here as a temperature dependency of the kinetic coefficients k and K_s (Figure 4 for acetate). O'Rourke (3), using the kinetic information in Table VI, developed the following equations of the form of Equation 20 to define the temperature dependency of k and K_c for a complex waste over the temperature range of 20°–35°C:

$$(k)_T = (6.67 \text{ day}^{-1})\ 10^{[-0.015(35-T)]} \tag{22}$$

$$(K_c)_T = (2224 \text{ mg/l COD})\ 10^{[0.046(35-T)]} \tag{23}$$

in which T = temperature (°C). Figure 7 shows the effect of temperature on effluent quality as a function of θ_c for a complex waste (Table VI and $Y = 0.05$ and $b = 0.01$). The principal effect of temperature is that treatment efficiency (effluent quality) at a given value of θ_c decreases with a decrease in temperature. Conversely a larger value of θ_c is required at a lower temperature to achieve a given level of treatment.

The designer should consider a number of alternatives when dealing with a medium or dilute strength waste at temperatures considerably below 35°C. These alternatives include heating the waste before treatment; attaining the desired value of θ_c by using the solids recycle process scheme; using an anaerobic filter to achieve very large values of θ_c. The design choice should be made on a least-cost basis. In proposing to heat the waste one must consider whether sufficient methane can be generated to effect the heating without need for costly fuel supplementation. McCarty (29) has stated that the waste COD must be equal to or greater than 5000 mg/liter to increase significantly the temperature of the influent waste without a supplemental fuel requirement.

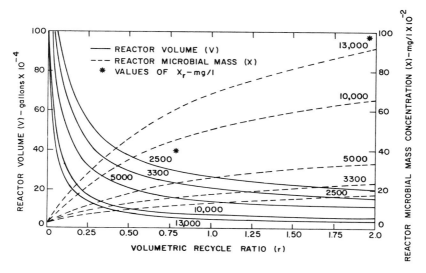

Figure 6.   Relationships between volumetric recycle ratio and reactor volume and microbial mass concentration for various values of X_r and an assumed specific efficiency of 97% in a hypothetical treatment situation

Non-Kinetic Factors. From the kinetic viewpoint there is no upper limit to the volumetric loading rate which may be imposed on an anaerobic biological process at a given θ_c if the assumptions of the steady state kinetic model are satisfied.   In actual practice the upper limit on the volumetric loading rate and hence volumetric waste stabilization rate for a given θ_c will be imposed by what may be called non-kinetic environmental factors.   For the conventional process these factors include (1) the mixing of concentrated sludges; (2) the provision of adequate nutrients for biological growth; (3) the elimination of toxic materials and the provision of a favorable ionic environment; and (4) the maintenance of a "neutral" pH.   In addition to factors 2 through 4, the performance of the solids recycle process is influenced by the efficiency of separation of the suspended solids from the mixed liquor and the rate of solids recycle. Several of these factors have been considered by McCarty (*29*) in a series of articles on the fundamentals of anaerobic treatment.   Any of these factors can cause deviations from the digester performance predicted by the kinetic model since the model assumes that substrate concentration is the only growth limiting factor.   Because "environmental factors" can affect digester performance adversely and because one of these factors will constitute the practical volumetric loading limit, every effort should be made to optimize these "environmental factors" in the final design of an anaerobic process.

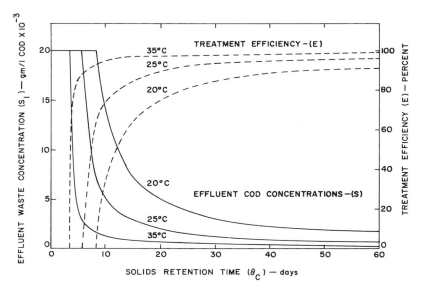

Figure 7. Predicted effect of fermentation temperature on effluent biodegradable COD concentration and treatment efficiency for a complex waste

Summary

The rate limiting step approach to process kinetics of anaerobic treatment provides the designer with a rational guide to the design of anaerobic treatment systems. Combining this approach with steady-state treatment models developed for biological processes, it is possible to design an anaerobic process to accomplish stated treatment objectives in terms of effluent quality or volumetric waste utilization. Six factors to be considered in selecting the design value of the principal design variable θ_c, biological solids retention time, have been described. Equations 5, 6, 10, 11, 12, and 13 form the basis of the suggested design model. The average values of Y and b in Table II and the values of k and K_s for acetate and propionate in Table III provide the basic design information for low lipid wastes. The values in Table VI and $Y = 0.04$ and $b = 0.015$ along with the temperature effect equations (22 and 23) provide the basic design information for high lipid content, complex wastes—*i.e.,* sewage sludge.

As knowledge of the microbiology and environmental requirements of methanogenic fermentations becomes more complete, it may be necessary to modify the specifics of the design approach suggested here. It is clear, however, that the process kinetic approach will continue to be the basis of a rational approach to biological waste treatment design.

Acknowledgments

The author's efforts were supported in part by the Cornell University Water Resources and Marine Sciences Center with Annual Allotment Funds (Project A-016-NY) provided by Office of Water Resources Research, United States Department of the Interior.

The author also acknowledges the contribution to this paper of the kinetic information on lipid containing complex wastes which appears in the unpublished doctoral dissertation of his friend and colleague, James T. O'Rourke (3).

Nomenclature

b = microorganism decay coefficient, time^{-1}

dF/dt = rate of microbial substrate utilization per unit volume, mass/volume-time

E = treatment efficiency, %

g = subscript which indicates performance evaluation based on a gross parameter—*e.g.*, BOD or COD

K_s = half velocity coefficient, equal to the substrate concentration when $(dF/dt)/X = (1/2)k$, mass/volume

K_c = composite half velocity coefficient for a complex waste, mass/volume

k = maximum rate of substrate utilization per unit weight of microorganisms, time^{-1}

P_x = excess microorganisms production rate, mass/time

Q = "net" flow rate of liquid through the reactor, volume/time

q = flow rate at which underflow of solids separator is recycled to the reactor, volume/time

r = volumetric recycle ratio—*i.e.*, q/Q

S = substrate concentration, mass/volume

s = subscript which indicates performance evaluation based on removal of a specific waste constituent

T = temperature

t = time

U = specific utilization defined as $(dF/dt)/X$, time^{-1}

V = reactor volume, volume

w = flow rate of that fraction of solids separator underflow which is wasted from the system, volume/time

X = microbial mass concentration, mass/volume

X_r = microbial mass concentration in the underflow from the solids separator, mass/volume

X_τ = total active microbial mass in treatment system, mass

Y = growth yield coefficient, mass/mass

0 = subscript which denotes reactor influent (raw waste) stream

1 = subscript which denotes reactor effluent stream

θ = reactor mean hydraulic retention time based on raw waste flow—*i.e.*, V/Q, time

θ_c = biological solids retention time as defined by Equation 4, time

$\theta_c{}^d$ = value of biological solids retention time used in design, time

$\theta_c{}^m$ = minimum biological solids retention time at which $S_1 = S_0$, time

$[\theta_c{}^m]_{\lim}$ = limiting value of $\theta_c{}^m$ which occurs when $S_0 >> K_s$, time

$\theta_c{}^*$ = value of θ_c at which volumetric waste utilization rate is a maximum (defined for conventional process only), time

μ = $(dX/dt)/X$—i.e., net specific growth rate of microorganisms, time^{-1}

ρ = rate of waste removal per unit volume of reactor, mass/volume-time

τ = subscript which indicates total system parameter

Literature Cited

(1) McCarty, P. L., "Anaerobic Treatment of Soluble Wastes," "Advances in Water Quality Improvement," E. F. Gloyna and W. W. Eckenfelder, Jr., Eds., pp. 336–352, University of Texas Press, Austin, 1968.

(2) Lawrence, A. W., McCarty, P. L., "Kinetics of Methane Fermentation in Anaerobic Treatment," *J. Water Pollut. Contr. Fed.* (1969) **41** (2) pt. 2, R1–R17.

(3) O'Rourke, J. T., "Kinetics of Anaerobic Treatment at Reduced Temperatures," Thesis presented to Stanford University, Stanford, Calif., in 1968, in partial fulfillment of the requirements for the degree of Doctor of Philosophy.

(4) Agardy, F. J., Cole, R. D., Pearson, E. A., "Kinetics and Activity Parameters of Anaerobic Fermentation Systems," *SERL Rept. 63-2* (Feb. 1963) Sanitary Engineering Research Laboratory, University of California, Berkeley.

(5) Andrews, J. F., Pearson, E. A., "Kinetics and Characteristics of Volatile Acid Production in Anaerobic Fermentation Processes," *Int. J. Air Water Pollut.* (1965) **9**, 439–461.

(6) Stewart, M. J., "Reaction Kinetics and Operational Parameters of Continuous-Flow Anaerobic-Fermentation Processes," *I.E.R. Ser. 90, Rept. 4* (June 1958) Sanitary Engineering Research Laboratory, University of California, Berkeley.

(7) McCarty, P. L., "Kinetics of Waste Assimilation in Anaerobic Treatment," "Developments in Industrial Microbiology," Vol. 7, pp. 144–155, American Institute of Biological Sciences, Washington, D. C., 1966.

(8) van Uden, N., "Transport-Limited Growth in the Chemostat and Its Competitive Inhibition; A Theoretical Treatment," *Arch. Mikrobiol.* (1967) **58**, 145–154.

(9) Monod, J., "La Technique of Culture Continue; Theorie et Applications," *Ann. Inst. Pasteur, Paris* (1950) **79**, 390–410.

(10) Novick, A., Szilard, L., "Experiments with the Chemostate on Spontaneous Mutations of Bacteria," *Proc. Nat. Acad. Sci. U.S.* (1950) **36**, 708–719.

(11) Lawrence, A. W., McCarty, P. L., "A Unified Basis for Biological Treatment Design and Operation," *J. Sanit. Eng. Div., Amer. Soc. Civil Eng.* (1970) **96** (SA3), 757–778.

(12) Pearson, E. A., "Kinetics of Biological Treatment," "Advances in Water Quality Improvement," E. F. Gloyna and W. W. Eckenfelder, Jr., Eds., pp. 381–394, University of Texas Press, Austin, 1968.

(13) McCarty, P. L., "The Methane Fermentation," "Principles and Applications in Aquatic Microbiology," H. Heukelekian and N. C. Dondero, Eds., pp. 314–343, Wiley, New York, 1964.
(14) Toerien, D. F., Hattingh, W. H. J., "Anaerobic Digestion, I. The Microbiology of Anaerobic Digestion," *Water Res.* (1969) **3**, 385–416.
(15) Bryant, M. P. *et al.*, "*Methanobacillus omelianskii*, a Symbiotic Association of Two Bacterial Species," *Bacteriol. Proc.* (1967) 19.
(16) Sawyer, C. N., Roy, H. K., "A Laboratory Evaluation of High-Rate Sludge Digestion," *Sewage Ind. Wastes* (1955) **27** (12), 1356–1363.
(17) Torpey, W. N., "Loading to Failure of a Pilot High-Rate Digester," *Sewage Ind. Waste* (1955) **27** (2), 121–133.
(18) Smith, P. H., Bordeaux, F. M., Shuba, P. J., "Methanogenesis in Sludge Digestion," "Abstracts of Papers," 159th National Meeting, ACS, Feb. 1970, WATR 49.
(19) Barker, H. A., "Bacterial Fermentations," Wiley, New York, 1956.
(20) Lawrence, A. W., McCarty, P. L., "Kinetics of Methane Fermentation in Anaerobic Waste Treatment," *Tech. Rept. 75* (Feb. 1967) Department of Civil Engineering, Stanford University, Stanford, Calif.
(21) McCarty, P. L., "Thermodynamics of Biological Synthesis and Growth," *Int. J. Air Water Pollut.* (1965) **9**, 621–639.
(22) Shea, T. G. *et al.*, "Kinetics of Hydrogen Assimilation in the Methane Fermentation," *Water Res.* (1968) **2**, 833–848.
(23) McCarty, P. L., "Energetics and Bacterial Growth," *Proc. Rudolfs Res. Conf., 5th*, Rutgers University, New Brunswick, N. J. July 2, 1969.
(24) Novak, S. T., Carlson, D. A., "The Kinetics of Anaerobic Long-Chain Fatty Acid Degradation," *Proc. Ann. Conf. Water Pollut. Control Fed., 42nd*, Dallas, Oct. 1969.
(25) Bryant, M. P., "Nutrient Requirements of Methanogenic Bacteria," ADVAN. CHEM. SER. (1971) **105**, 23.
(26) Pfeffer, J. T., "Increased Loadings on Digesters with Recycle of Digested Solids," *J. Water Pollut. Control Fed.* (1968) **40** (11), pt. 1, 1920–1933.
(27) Torpey, W. N., Melbinger, N. R., "Reduction of Digested Sludge Volume by Controlled Recirculation," *J. Water Pollut. Control Fed.* (1967) **39** (9), 1464–1474.
(28) Young, J. C., McCarty, P. L., "The Anaerobic Filter for Waste Treatment," *Tech. Rept. 87* (March 1968) Department of Civil Engineering, Stanford University, Stanford, Calif.
(29) McCarty, P. L., "Anaerobic Waste Treatment Fundamentals: I. Chemistry and Microbiology; II. Environmental Requirements and Control; III. Toxic Materials and Their Control; IV. Process Design," *Pub. Works* (1964) **95** (9), 107–112; (10), 123–126; (11), 91–94; (12), 95–99.

RECEIVED August 3, 1970.

INDEX

INDEX